GENDER EQUALITY AND WATER SECURITY

A CONCEPTUAL FRAMEWORK AND PRACTICAL STRATEGIES TO ACCELERATE PROGRESS

OCTOBER 2024

ASIAN DEVELOPMENT BANK

6 ADB Avenue, Mandaluyong City, 1550 Metro Manila, Philippines
Tel +63 2 8632 4444; Fax +63 2 8636 2444
www.adb.org

ISBN 978-92-9270-926-6 (print); 978-92-9270-927-3 (PDF); 978-92-9270-928-0 (ebook)
Publication Stock No. TCS240469-2
DOI: http://dx.doi.org/10.22617/TCS240469-2

Notes:
In this publication, "$" refers to United States dollars.
ADB recognizes "Vietnam" as Viet Nam.

Cover design by Ginojesu D. Pascua.

Front cover photo: Women carrying water in Rajasthan, India. Around the world, women are often responsible for carrying water from sources to the home (photo by gnomeandi – stock.adobe.com).

 Printed on recycled paper

Contents

Tables, Figures, and Boxes

Tables

Figures

Boxes

Acknowledgments

The work was guided by Director Neeta Pokhrel and Principal Water Security Specialist Allison Woodruff of the Water and Urban Development Sector Office, Sectors Group of the Asian Development Bank (ADB). ADB's Water Inclusion Advisory Group commented on an earlier version of the conceptual framework and reviewed this report. This publication was prepared by Juliet Willets, Jessica MacArthur, and Melita Grant from the University of Technology Sydney's Institute for Sustainable Futures.

The work also benefited from contributions and review by the participants of the Inclusion Roundtable.

ADB Inclusion Advisory Group

Neeta Pokhrel
Allison Woodruff
Silvia Cardascia
Wei Kim Swain
Sunila Ghimire
Siti Hasanah
Hisaka Kimura
Rosemary Victoria Atabug
Marjana Chowdhury
Yasmin Siddiqi
Yukiko Ito
Malika Shagazatova

Inclusion Roundtable Members

Jaehyang So, Global Water Partnership
Rachael McDonnell, International Water Management Institute
Ryuji Ogata, Japan International Cooperation Agency
Claudine Orocio-Isorena, Metropolitan Waterworks and Sewerage System
Alejandro Jimenez, Stockholm International Water Institute
Yumiko Noda, Veolia Japan
Laura Veronica Imburgia, United Nations Educational, Scientific and Cultural Organization (UNESCO)
Shreya Gyawali, Australian Water Partnership
Muyatwa Sitali, Sanitation and Water for All
Sera Young, Northwestern University

Executive Summary

Why consider women's water security? There is significant evidence that while women experience greater impacts of water insecurity than men, they are underrepresented in formal and informal water management and service delivery institutions. These differences arise from broader gender inequalities and marginalization of particular groups within the relevant societies.

People experience water security and insecurity differently. Women are often responsible for water collection and management within households. Women and girls are also more likely to experience food insecurity; personal safety concerns; vulnerability to disasters, severe weather events, and climate change; risks of disease and physical and mental health impacts; and reduced economic opportunities related to water. It is therefore important to consider the specific factors that contribute to women's water insecurity. To achieve water security for all, it is imperative that gender and inclusion are meaningfully incorporated into policies, governance, and management processes.

This report presents findings from a collaborative knowledge work among the Asian Development Bank; University of Technology Sydney's Institute for Sustainable Futures; and a wider network of global partners interested in progressing the gender and inclusion agenda in water security. This knowledge work aims to (i) develop a conceptual framework to guide a gendered perspective on water security in policy, project design, analysis, and monitoring; and (ii) document practices on how women's inclusion and gender equality in the water sector can be enhanced toward more transformational gendered approaches.

Conceptual framework for gendered water security. The conceptual framework adopts a human-centric perspective and is designed to highlight the societal, institutional, environmental, and biophysical contexts that impact diverse individuals' experience of water security and insecurity (Figure 1). While focused on gender equality and the experiences of women, the framework also incorporates an intersection lens covering diverse individual characteristics and can be helpful in considering other vulnerable groups. The framework is intended to help water professionals identify critical factors in conceptualizing and measuring gendered water security across domestic, productive, and cultural water uses as well as in relation to water-related disasters. This, in turn, can support improved gender equality and inclusion in policy development and project design, as well as in monitoring and evaluation efforts.

Transformative practices to support enhanced inclusion and gender equality. This report also provides 33 short cases to demonstrate how gender considerations can be better integrated into water project design. Using a second framework presented in Figure 2, these cases highlight the spectrum of possible entry points for promoting improved gender equality and inclusion, ranging from gender-sensitive, gender-responsive, to gender-transformative approaches.

The case examples cover four key areas relevant to water security: (i) enhancing women's water and sanitation access, (ii) reducing women's vulnerability to water-related disasters, (iii) improving equitable water management and access to irrigation water, and (iv) strengthening women's employment in water organizations. Across all areas, accompanying "do no harm" strategies are needed to address and mitigate resistance and backlash. This report recommends adopting five broad strategies, as illustrated by the case examples:

(i) **Embed gender thinking in policy considerations.** Doing so reflects the significance of gender within issues of water security. This includes work to strengthen policies for inclusive infrastructure design (case 3), urban master plans (case 6), disaster early warning systems (cases 13 and 14), roles in water use committees and clubs (cases 18 and 21), and utility policies (cases 24 and 25).

(ii) **Promote meaningful engagement, promote codesign and participation, and address barriers.** Promote meaningful engagement of women, girls, and other marginalized individuals, and avoid harmful token participation. This can include the participatory design of infrastructure, products, standards, programs, and services (cases 1, 2, 3, 4, and 23), participatory implementation (cases 5, 7, 9, 12, 15, 16, 17, 18, 19, 20, 21, and 24), and participatory monitoring and evaluation (cases 8 and 13).

(iii) **Support and increase women's voice, leadership, and decision-making in water and sanitation service provision and wider water management.** Doing so empowers women working and volunteering in water and sanitation service provision, including in the private sector (cases 22, 31, 32, and 33), the public sector (case 25), civil society (case 28), and communities (cases 5, 16, 17, and 18).

(iv) **Address systemic inequality within organizations.** This can be done within formal and informal water-related organizations through analysis (cases 29 and 30), dialogue and training such as unconscious bias (cases 14 and 26), strategic partnerships (case 10), and celebration of champions (cases 27 and 28).

(v) **Monitor, research, and evaluate gendered outcomes and impacts, both intended and unintended.** It is important to regularly track the impacts of water projects, as well as policies and institutional arragements, to identify issues early and allow for course corrections. Monitoring processes should include and go beyond sex-disaggregated data by taking into account intersectional aspects, practical gender needs, and women's strategic gender interests. This requires a foundation of strong gender analysis (cases 4 and 6) to look at individual impacts (cases 8, 11, and 13) and institutional trends (case 29), as well as to pay attention to unintended outcomes.

In summary, the recognition of gender and inclusion in water security conceptualizations, implementation strategies, and monitoring efforts are critical for ensuring water security. Without attention to these

Women and water. To achieve water security for all, it is imperative that gender and inclusion are meaningfully incorporated into policies, governance, and management processes (photo by Amit Verma/ADB).

dimensions, efforts will fall short and continue to result in unequal benefits and suboptimal water security and societal outcomes. Equally, with intention, forethought, and willingness to challenge existing norms, significant progress to reduce inequalities is possible.

Addressing gender dynamics in water management has the potential to accelerate equitable water security both in Asia and the Pacific and globally. Addressing gender equality with an intersectional lens will ensure other dimensions of marginalization and inclusion are also given attention.

The four key recommendations of this report are to (i) build internal capacity to address gender equality in water-related international agencies and development partners, (ii) address gender equality across institutions responsible for governance and management, (iii) set targets for and monitor progress on gender equality in water security, and (iv) promote a shift toward gender-transformative practice in water security initiatives.

I. Background

To achieve water security for all, it is imperative that gender and inclusion are meaningfully incorporated into policies, governance, and management processes. Water insecurity disproportionately impacts women and girls. Women and girls are also more likely to experience food insecurity, personal safety concerns, vulnerability to disasters, risks of disease and physical and mental health impacts, and reduced economic opportunities related to water. Box 1 highlights the critical intersections between gender and water and the unique challenges faced by women and girls in relation to water.

Women's inclusion and gender equality need to be enhanced in all parts of water management. Their shared involvement includes participation in management of water resources; design, planning, and delivery of water and sanitation services in both formal and informal roles; water-sharing arrangements; and strategies to reduce water-related risks. Addressing gender dynamics in water management can accelerate equitable water security in Asia and the Pacific and globally. Addressing gender equality with an intersectional lens will ensure other dimensions of marginalization and inclusion are also given attention.

The gender strategy of the Asian Development Bank (ADB) for 2019–2024 commits to scale up gender mainstreaming in operations across sectors and to integrate a transformative gender agenda (ADB 2019); and one of the five guiding principles of ADB's *Water Sector Directional Guide 2030* is "promoting inclusiveness and gender equality" (ADB 2022). While ADB's *Asian Water Development Outlook* series provides an established approach to measuring water security in Asia and the Pacific, this report provides the basis for evolving that approach to better differentiate water security with respect to gender or inclusion.

Box 1: Gender and Water at a Glance

- Globally, 2 billion people lack access to safely managed water services and 3.6 billion people lack access to safely managed sanitation services (JMP 2021). Access to water and sanitation reflects significant differences between rural and urban populations and across wealth categories (JMP 2019).
- In 7 out of 10 households, women and girls are primarily responsible for water collection (JMP 2023) and each day, women and girls spend 200 million hours collecting water around the world (UNICEF 2016).
- Women represent only 18% of workers within water utilities (World Bank 2019) and 17% of the water and sanitation workforce more broadly (IWA 2014). Women comprise only 7% of graduates in engineering and construction (World Economic Forum 2022).
- A survey of 117 transboundary river basin organizations worldwide found that women made up fewer than one-third of their staff and fewer than one-fifth of staff in the highest leadership positions (Best 2019).
- While 80% of the world's food is produced by small-scale farming, and women make up on average 43% of this agriculture labor in developing countries, the benefits of irrigation do not accrue equally to men and women (FAO 2021; Pearl-Martinez 2017).

II. Rationale—Why Consider Women's Water Security?

There is significant evidence that women experience greater impacts of water insecurity than men and are underrepresented in formal and informal water management and service delivery institutions. These differences arise from broader gender inequalities and marginalization of particular groups within the relevant societies. There are six main reasons why women and water security is important:

(i) **Water insecurity has stronger impacts on women than on men.** For instance, in male-dominated cultures, women take the workload of collecting water (Dickin and Caretta 2022; JMP 2023). Water insecurity also brings negative health impacts on women (Grasham, Korzenevica, and Charles 2019). Women also experience worse impacts from water-related disasters (Jagnoor et al. 2019; Moayed et al. 2020) and higher risks of violence (Nunbogu and Elliott 2022). Further, households headed by women are more vulnerable to climate change (Alhassan, Kuwornu, and Osei-Asare 2019).

(ii) **Women can benefit significantly when water security is improved.** Evidence points to improved economic, physical, psychological, and relational well-being of women and girls when water security is improved (Winter, Darmstadt, and Davis 2021). These improvements are stronger for women, girls, and other marginalized groups than men (Caruso et al. 2022; De Guzman et al. 2023), and therefore are socially differentiated.

(iii) **Women are integrally connected to the management and stewardship of water.** Globally, in 80% of water-deprived households, women and girls are the primary collectors of water for domestic uses (UN Women 2018). Each day, women and girls spend 200 million hours collecting water around the world (UNICEF 2016). Additionally, in many societies, women hold a rich spiritual connection with water as water stewards (Awume, Patrick, and Baijius 2020).

(iv) **Women's expertise and knowledge with respect to water security are underutilized.** Women are underrepresented in leadership roles in river basin and water resources management organizations (Sehring, ter Horst, and Zwarteveen 2022) and utilities (World Bank 2019). They are also underrepresented in private sector and entrepreneurship roles, for instance, in water and sanitation businesses (Grant, Willetts, and Huggett 2019; Indarti et al. 2019; Soeters et al., 2020). However, evidence shows that the involvement of women can lead to improved functionality of water systems (Mommen, Humphries-Waa, and Gwavuya 2017).

(v) **The water security field has so far maintained a biophysical, material focus that is depoliticized and overlooks gender inequalities.** A new thinking based on feminist theory, political economy, and political ecology offers pathways to address this deficiency and give attention to socially differentiated experiences that underpin water insecurities (Truelove 2019). Experts emphasize the importance of applying the human capabilities approach to water security (Jepson et al. 2017a), and linking gender, poverty, and discrimination to water security inequalities, particularly in the face of climate change (Fröhlich et al. 2018; Grasham, Korzenevica, and Charles 2019).

Studies and tools for gendered aspects of water security are limited. Octavianti and Staddon's (2021) recent review of water security assessment tools did not include any reflections on gender. Reviews also have noted the absence of evidence, for instance, on gendered inequalities in water security in Sub-Saharan Africa (Grasham, Korzenevica, and Charles 2019) and the lack of interlinkages between gender equality (Sustainable Development Goal 5) and urban water security (Grant, Willetts, and Huggett 2019; Tandon et al. 2022).

A review of policy keywords across 58 national water resources management and water and sanitation revealed that policies and laws in Asia and the Pacific have limited mention of gender and equity, and no mention of equality nor inclusion (Figure 1). Policy keywords were identified in the FAOLEX Database of the Food and Agriculture Organization of the United Nations (FAO); and the most common keywords are water supply, potable water, water resource management, and policy and/or planning.[1]

Addressing these inequalities, including access to safely managed water supply and sanitation and water resources, as well as ensuring that women are well represented as water leaders, policymakers, and professionals, are essential steps to accelerate progress on water security and drive resilience in the face of climate change.

Figure 1: Coverage of Gender and Inclusion in Policy Keywords

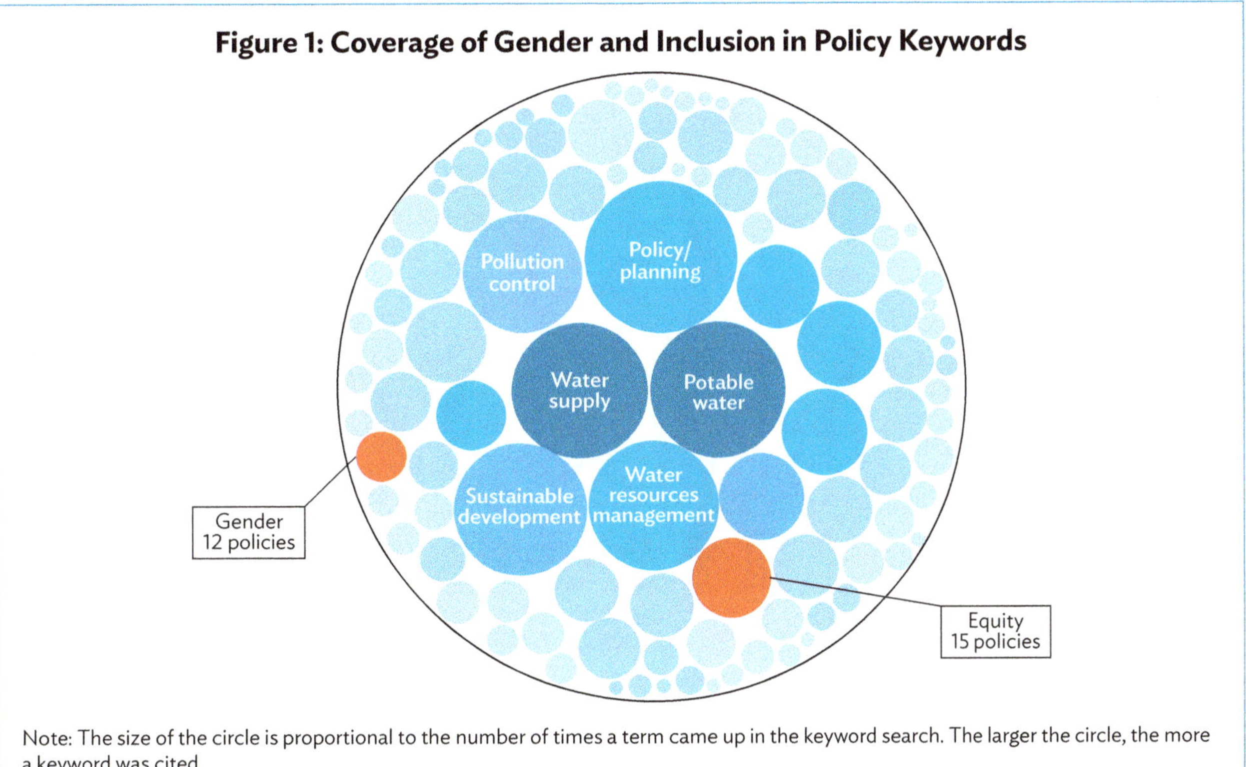

Note: The size of the circle is proportional to the number of times a term came up in the keyword search. The larger the circle, the more a keyword was cited.

Source: University of Technology Sydney's Institute for Sustainable Futures.

[1] FAOLEX is a database of national laws, regulations, and policies related to food, agriculture, and natural resources management. The database also includes indexing information such as abstracts and keywords compiled by the FAO.

Water labor. Women are often responsible for water collection and management within households (photo by Ariel Javellana/ADB).

III. Conceptual Framework for Gendered Water Security

Why a new framework? Most historical frameworks for water security adopt a macro perspective integrating national or basin-level water security and foregrounding water resources (Butte et al. 2022). More recent frameworks include a focus on households, taking a human-centric capability approach (Jepson et al. 2017a). However, to date, there is limited conceptualization of gendered water security that addresses socially differentiated individual and intra-household experiences. Since conceptualizations translate into ideas about aspects to plan, monitor, measure, and evaluate water security (Octavianti and Staddon 2021), there is a need for a broader, socially differentiated view to ensure that gendered dimensions are given attention.

This chapter presents a conceptual framework for gendered water security (Figures 2 and 3). The framework identifies the conditions through which socially differentiated water insecurity may manifest, and its subsequent gendered impacts on well-being.

This conceptual framework defines important dimensions for planning and assessing gender and inclusion aspects of water security. It will be useful to water professionals in program design and measurement, as well as at the policy level where it can inform the breadth of areas where targets may be set and where progress can be tracked. For example, the following scenarios are possible:

Figure 2: Gendered Water Security Framework

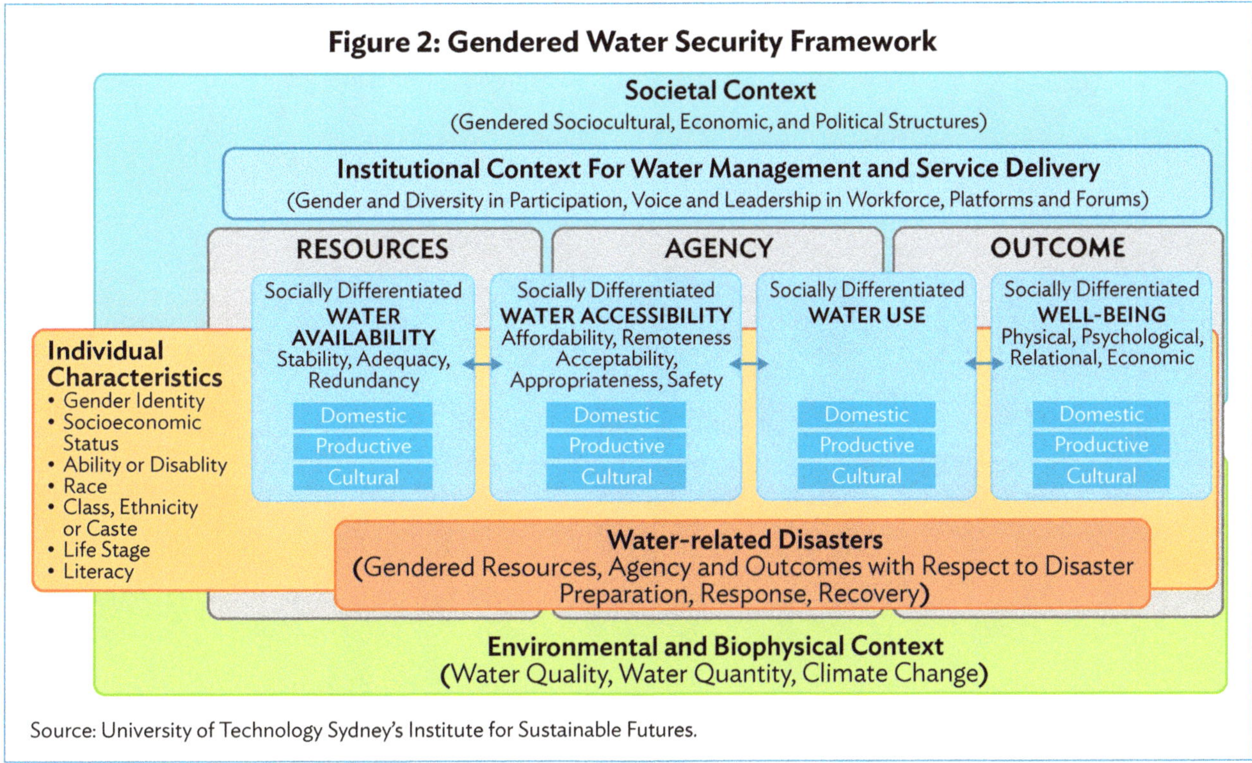

Source: University of Technology Sydney's Institute for Sustainable Futures.

- **During program design.** The framework can be used to identify key areas where a program has important gendered aspects that need consideration, whether in relation to the societal and institutional context; gendered aspects of water availability, accessibility, use, and well-being outcomes in relation to domestic, productive, or cultural purposes; or gendered aspects of water-related disasters.
- **Defining scope and dimensions of gender analysis.** Since the framework covers the full extent of possible gendered aspects, a subset of framework aspects that are most relevant to a given program can be chosen for further analysis.
- **Informing monitoring and evaluation.** The framework provides important ideas for monitoring and evaluation, particularly with respect to unintended outcomes, ensuring that monitoring and evaluation captures the experiences and outcomes for diverse people.

The framework is informed by gender and development theory and the work of Kabeer (1999; 2018), who identifies resources, agency, and outcomes in the context of "structures of constraint" as a way to consider human capabilities and achievements (Figures 2 and 3). These gendered aspects are overlayed with common dimensions of water security (availability, accessibility, and water use for different purposes) and resultant well-being—all of which are contingent on an individual's characteristics within the given societal and institutional context of the water sector—and are thus "socially differentiated." Similarly, the gendered experiences of preparing and responding to water-related disasters are shaped by the societal and institutional context and individual characteristics and relative access to resources and agency with differentiated outcomes.

Next, we describe each component of the gendered water security framework and provide justification for a socially differentiated and intra-household gendered perspective, drawing on academic literature, including review papers and specific studies on aspects of water security.

Women carry water on their heads in Jaisalmer, India (photo by TMAX - stock.adobe.com).

Gender Equality

This framework incorporates three important dimensions of women's empowerment and the capability approaches' theory of gender equality for human development (Kabeer 1999; Nussbaum 2000). Recent iterations of Kabeer's (1999) framework highlight how structures of constraint and freedom influence equality (Robeyns 2005), and these "structures" are described below under "societal context." This individual-centric framework includes the resources, agency, and outcomes that support the flourishing of individuals and therefore society:

- **Resources.** Access to, use of, and control of the tangible and intangible materials, and human and social resources. These resources are gained and exchanged through social relationships (Kabeer 1999). In this framework, the relevant resources include the availability of the water resource at required water quality to relevant individuals and its accessibility, as well as resources that offer individuals protection from the impacts of water-related disasters.
- **Agency.** The ability to define goals and act upon them through processes of decision-making, bargaining and negotiation, and resistance (Kabeer 1999). In this framework, agency pertains to how women, men, and other genders access and use water in different ways (including for personal, domestic, and productive purposes) and how their agency effectively prepares for, responds to, and recovers from water-related disasters.
- **Outcomes.** Also known as achievements, the multidimensional outcomes of individual human flourishing or well-being (Kabeer 1999; Robeyns 2005). In this framework, the outcomes are well-being outcomes, including well-being in terms of health and economic prosperity.

Women are collecting water. ADB-supported initiatives help provide access to clean water supply (photo by Samir Jung Thapa/ADB).

Figure 3: Gendered Water Security Framework—Explanation of Dimensions

Societal Context

Gendered structures that impact how water is governed, managed, accessed, and used by different individuals.

- **Sociocultural** (taboos, social norms, beliefs, and customs)
- **Political** (rights, citizenship, and land/natural resource access)
- **Economic** (financial dependencies, livehoods, and employment)

Individual Characteristics

The intersectional characteristics that shape patterns of insecurities and inequalities, based on the societal and institutional context. This includes aspects of gender-identity, class, race, religion, ethnicity, life stage/age, ability/disability, socioeconomic status, literacy, occupational status, and residential status.

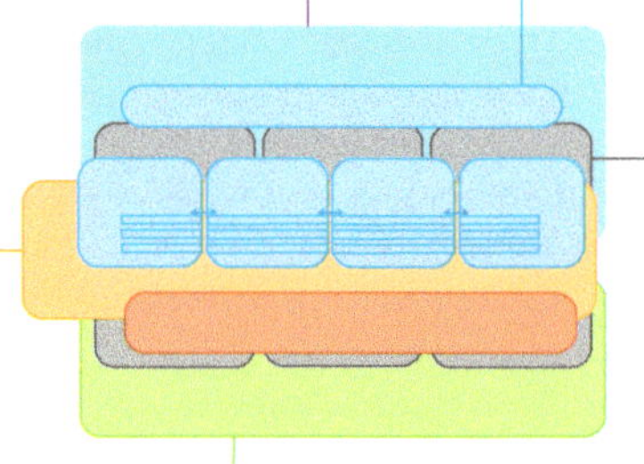

Institutional Context

Water management and governance shape diversity in the workforce and informal institutions, including for basins and transboundary water governance, irrigation, as well as water and sanitation service delivery.

- **Participation** (meaningful inclusion paid and unpaid in workforce, platform, forums)
- **Voice** (rights, citizenship, and land/natural resource access)
- **Leadership** (financial dependencies, livehoods, and employment)

Gender Equality

An individual and human development approach to equality and empowerment.

- **Resources** (gendered access to, use of, and control of tangible and intangible support preparedness for water-related disasters)
- **Agency** (gendered ability and confidence to define goals and act upon them through decision making and negotiation)
- **Outcomes** (gendered achievement of human flourishing and well-being)

Environmental and Biophysical Context

The water cycle, including surface and groundwater, and their respective water quality and quantity. This environmental context and hydrological cycle is affected by climate, including trends and events and water-related disasters. The dimension of the framework is not gendered, as it represents the physical aspect.

Socially Differentiated Water Availability

Beyond the environmental and biophysical context and shaped by the societal context:

- Stability (reliable availability of consistent water without seasonal or other disruption)
- Adequacy (availability of enough water to meet diverse water needs of different individuals)
- Redundancy (availability of multiple sources of water supporting individuals and their households to better absorb economic, political, and climatic shocks)

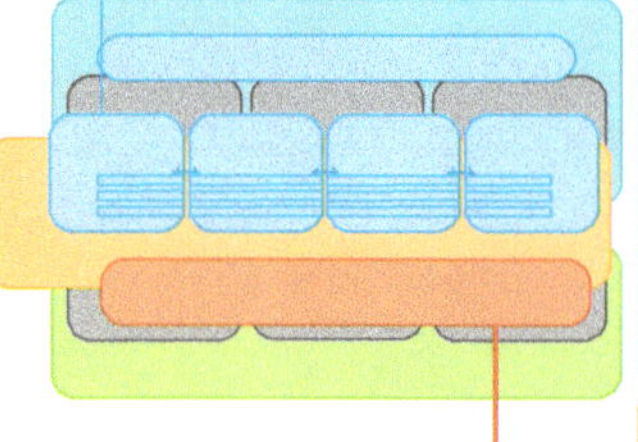

Socially Differentiated Water Accessibility

Shaped by societal context, access through diverse mediums such as wells, pipes, taps, and irrigation:

- **Affordability** (ability to pay for water in ways that do not compromise other personal or household resource needs, differentiated willingness to pay)
- **Remoteness** (individual experience of distance of acceptable water sources to the home and the opportunity cost of time spent on water collection)
- **Acceptability** (individual perceptions of water quality, color, smell, and taste)
- **Appropriateness** (acceptable water, sanitation, irrigation, and other technologies and their usability by all genders)
- **Safety** (personal safety while acquiring water or during defecation and the safety of water for use and consumption)

Socially Differentiated Water Use

Diverse and seasonal priorities and patterns of water use and reuse.

- **Personal** (personal hygiene for washing, bathing, defecation, menstrual hygiene and drinking)
- **Domestic** (household-related chores often gendered including care work, cooking, and amenities)
- **Productive** (livelihood uses related to gendered roles in irrigation, agriculture, livestock rearing, gardens, and other income-earning activities)
- **Cultural** (gendered spiritual, ceremonial, and recreational uses)
- **Wastewater exposure and reuse** (differential exposure to risk or benefit)

Socially Differentiated Well-being Impacts

Positive or negative gendered impacts on personal well-being and flourishing.

- **Physical** (impacts through food security, nutrition, water quality, water carrying, and pollution)
- **Psychological** (impacts on mental health through shame, fear, anger, frustration, embarrassment, negative identity and marginalization; or positive mental health outcomes)
- **Relational** (interactions with other individual, conflict and violence or harmony)
- **Economic** (income, financial security, and educational impacts)

Socially Differentiated Water Accessibility

Gendered resource, agency, and outcomes related to experiences of water-related disasters.

- **Preparation** (relative opportunity to access skills and resources and make decision to prepare for disasters, including differential access to early warning systems, communications, swimming skills, etc.)
- **Response** (differences in mobility, decision making authority, and ability to meet individual needs potentially leading to compromised safety and exposure to violence)
- **Recovery** (extent of pyschological trauma, stress, and relative access to resources and decision making to recover and improve resilience)

Source: University of Technology Sydney's Institute for Sustainable Futures.

Societal Context

The "structures of constraint and freedom" described in the gender and development field in this framework are represented as "societal context." The societal context contains the economic, sociocultural, and political environments that produce and sustain gender roles, norms, and dynamics, underpinned by power dynamics. These structures can include institutions and socialization processes in the home, workplace, and broader society.

- **Sociocultural.** The traditions, norms, beliefs, and customs that shape the interrelations between social structures and patterns of water security (Crow and Sultana 2002; Fröhlich et al. 2018). This includes the patriarchal gendered division of labor that compels women to be primarily responsible for household-level water and sanitation tasks and men to be primarily responsible for community-level water and sanitation tasks, which is particularly prevalent in rural agrarian communities (Dickin and Caretta 2022; Tantoh et al. 2021). Cultural identity also shapes the stewardship of water resources, and in some cultures, women remain the stewards of water resources (Awume, Patrick, and Baijius 2020). In other sociocultural systems, women are made more vulnerable to water-related disasters because of sociocultural norms, taboos, and beliefs related to mobility and skill sets such as swimming (Moayed et al. 2020).
- **Political.** The legal dimensions that shape water security include rights, citizenship, and land access. Truelove (2019) describes this as "everyday practices and politics, in relation to both governance and citizens." Additionally, limited legal access to resources and restricted formal rights, such as land tenure, result in a lack of access and control of resources by women and other marginalized groups (Imburgia 2019; Tantoh et al. 2021). For example, on the Texas-Mexico border, immigration status affected water security more than any other socioeconomic factors (Jepson and Vandewalle 2016). Women are also often politically restricted from participating in decision-making forums for water management (Imburgia 2019; Leya 2019). In many contexts, political dimensions lead women to have higher vulnerability during water-related disasters, including restricted access to resources and support, which can lead to greater risk of violence and abuse (Moayed et al. 2020).
- **Economic.** The formal and informal dimensions that shape economic opportunities, such as financial dependencies, livelihoods, and employment (De Guzman et al. 2023; Grasham, Korzenevica, and Charles 2019). At the macro level, the water sector demonstrates weak gender parity, with men being dominant in water utilities, particularly in leadership roles (World Bank 2019), and with women facing barriers as entrepreneurs (Grant et al. 2019; Indarti et al. 2019; Rautanen and Baaniya 2008) and in wider water management organizations (Cleaver 1998; Grant et al. 2017; van Wijk-Sijbesma 1997). At the micro level, economic systems shape the roles and responsibilities for household financial security and women's financial independence (De Guzman et al. 2023). There are also strong interactions between poverty and water security, exacerbated by climate change (Grasham, Korzenevica, and Charles 2019; Nunbogu and Elliott 2022).

Institutional Context

Embedded within the societal context are the formal and informal institutions and systems that manage water resources and deliver water services (including for domestic water, sanitation, and hygiene needs; for irrigation; and for broader agriculture, income-generation, or industrial purposes) and that are influenced by sociocultural, political, and economic structures.

- **Participation.** The meaningful inclusion, both paid and unpaid, of diverse individuals that manage and deliver water services in platforms and forums (Agarwal 2001; Cornwall 2003; Girard 2014). Paid participation of women in the management and delivery of water remains weak in utilities (World Bank 2019) and in more localized entrepreneurship (Grant et al. 2019; Indarti et al. 2019; Rautanen and Baaniya 2008). In some cases, where men are paid for their participation, women are expected to do the work as volunteers (van Wijk-Sijbesma 1997). Whether paid or unpaid, women's participation is often at risk of being tokenistic (Das 2014; O'Reilly 2004), but their participation has also been shown to improve water system performance (Mommen, Humphries-Waa, and Gwavuya 2017).
- **Voice.** The ability for diverse individuals to meaningfully influence the management and delivery of water services (Cornwall 2003; Dickin and Caretta 2022). In this aspect, it is not merely enough that women are invited to the table, but it is important that their voices are also heard in the management and delivery of water and sanitation (Dickin and Caretta 2022; Megaw and Winterford 2020; van Koppen et al. 2021).
- **Leadership.** The meaningful representation of diverse individuals in positions of decision-making authority and influence within platforms and forums that manage and deliver water services (Agarwal 2001; Cornwall 2003). In the water and sanitation sectors, women and other marginalized groups are underrepresented in formal positions of sector leadership and decision-making (World Bank 2019; Worsham 2021).

Water innovation. A hand pump and water containers in rural India (photo by Ecoview - stock.adobe.com).

Environmental and Biophysical Context

The overall environmental and biophysical context concerns the water cycle, including surface water and groundwater and their respective water quality and quantity. This dimension of the framework is not gendered, as it represents the physical reality of the water resource in its environmental context. A key driver in this context is climate change, which is creating major changes in the hydrological cycle (IPCC 2023).

Individual Characteristics

Individual water security is underpinned by the intersectional characteristics that shape patterns of insecurities and inequalities (Myrttinen 2017; Narain and Roth 2002; Soeters et al. 2019; Thompson 2016; Truelove 2019). This includes intersectional aspects of gender identity, class, race, religion, ethnicity, life stage and/or age, ability and/or disability, socioeconomic status, literacy, occupational status, and residential status (Collins 2015).

Socially Differentiated Water Availability

Multiple authors argue that the availability of water is socially differentiated, and goes beyond the environmental and biophysical context, as it is also deeply connected to societal factors stated in Tallman et al. (2022). This is true for water for domestic, productive, and cultural purposes. For example, poverty and immigration status can directly influence the availability of stable and reliable water (Jepson et al. 2017b; Jepson and Vandewalle 2016). Such intersectional characteristics interact with gender dimensions and result in social differentiation in terms of the following three areas:

- **Stability.** The reliable availability of water consistently without seasonality or disruptions (Elliott et al. 2019; Young et al. 2021a). Stable and reliable availability can be met through multiple sources or through sources that are more able to overcome seasonal constraints such as boreholes (Elliott et al. 2019). However, the stability of water sources is also related to increasing demand, overextraction, and climate change (Jain et al. 2021). Poor and marginalized individuals are often more at risk of having unstable water access influenced primarily by economic and political factors (Jepson and Vandewalle 2016).
- **Adequacy.** The availability of sufficient water to meet diverse water needs (Elliott et al. 2019; Narain and Roth 2002; Thompson et al. 2001). This principle highlights the value of thinking beyond drinking water to also include water for cleaning, household maintenance, construction, livestock, and irrigation (Elliott et al. 2019; van Koppen et al. 2021). Poor and marginalized individuals are often at risk of only getting enough water to meet basic needs, usually provided through subsidy programming (Thompson et al. 2001).
- **Redundancy.** The availability of multiple sources of water supporting households to better absorb economic, political, and climatic shocks (Elliott et al. 2019; Thompson et al. 2001). The availability of multiple sources has been shown to improve water security for poor people (Cleaver 1998; Elliott et al. 2019).

Fetching water. Water collection using a wheelbarrow in arid West Africa (photo by stock.adobe.com).

Socially Differentiated Water Accessibility

Once water is available, it is then important to explore the socially differentiated accessibility of the sources obtained through various sources, such as taps, wells, manual water collection, vendors, and pipes. Water accessibility is influenced by social, cultural, economic, and political aspects (Crow and Sultana 2002; Jaren, Leya, and Mondal 2022; Young et al. 2021a), and across domestic, productive, and cultural purposes.

- **Affordability.** The "ability to pay for water in ways that do not compromise other household resource needs" (Tallman et al. 2022, 2). Financial affordability relates to relevant tariffs, and can include capital expenditures, operation and maintenance, and pay-for-use.
- **Remoteness.** This is related to the distance of acceptable water sources to the home and the opportunity cost of time spent on water collection (Pouramin, Nagabhatla, and Miletto 2020). Irrigation programs that do not consider women and marginalized people's access to land and land tenure arrangements may end up excluding women and other vulnerable groups (Theis et al. 2016).
- **Acceptability.** The socially differentiated acceptability of the perceived water quality, including aspects such as color, taste, and perceived quality, which can vary by gender (Crow and Sultana 2002; Genter et al. 2023).
- **Appropriate technology.** From a technical perspective, water and sanitation technologies themselves must also be acceptable and usable by all genders (Elmendorf and Buckles 1980; UNICEF, WaterAid, and Water and Sanitation for the Urban Poor 2018). For instance, irrigation treadle pumps in Zambia have not been effective, as they were deemed inappropriate for women to use (iDE 2020). There have been assumptions that transferring irrigation technologies to women will give them control. However, this was not always the case due to gender dynamics at the household, intra-household, and community levels (Theis et al. 2016).
- **Safety.** "Both personal safety while acquiring water and the safety of water for use and consumption" (Tallman et al. 2022, 2). Firstly, safety in accessing water is socially differentiated. Around the world, women and children are the primary collectors of water and are often at risk of physical, sexual, psychological, and structural violence (Bisung and Elliott 2017; Nunbogu and Elliott 2022; Tallman et al. 2022). This can also include the sexual extortion of women while accessing water (Merkle, Allakulov, and Gonzalez 2022; United Nations Development Programme—Stockholm International Water Institute Water Governance Facility 2017). Secondly, vulnerable populations are more at risk of having poorer quality water with contamination such as salinity, arsenic, and fecal matter (Crow and Sultana 2002; Genter et al. 2023).

Socially Differentiated Water Use

Once water is accessible, individuals also have socially differentiated priorities for water use and different seasonal use patterns (Cleaver 1998; Crow and Sultana 2002; Varua et al. 2018). This includes personal, domestic, productive (economic), and cultural uses (Crow and Sultana 2002; van Koppen et al. 2021).

- **Personal.** Personal uses of water include aspects of personal hygiene (washing, bathing, and menstrual hygiene) and water for drinking (Thompson et al. 2001). These are socially differentiated and vary across life stages.
- **Domestic.** Domestic uses of water center around the running of a household and are strongly gendered (Cleaver 1998; Crow and Sultana 2002). Women are traditionally responsible for most domestic (sometimes referred to as "reproductive") activities such as cooking; washing dishes; cleaning; caring for subsistence gardens and livestock; and caring for children, the sick, and older people (Bryan and Lefore 2021; Thompson et al. 2001). Men are noted in some cultures as using water not only within a homestead but also outside the physical home for activities such as construction, washing of vehicles, and watering of lawns—referred to as "amenity" uses in other literature (Thompson et al. 2001).
- **Productive.** "Productive" or "economic" uses of water include livelihood and income-generating activities related to irrigation, agriculture, livestock raising, beer brewing, and income-earning gardens (Thompson et al. 2001). Women derive more of their income from productive water use than men (van Hoeve and van Koppen 2005; van Houweling et al. 2012; Marks et al. 2016). As this framework focuses on individual scale, wide-scale industrial use is not included, but only localized individual uses. Notably, the "gendered patterns of participations and benefits" from small-scale irrigation are not guaranteed and require more research (Bryan and Lefore 2021).
- **Cultural.** Often overlooked, cultural uses of water include spiritual, ceremonial, and recreational purposes (Baker et al. 2015; Jepson and Vandewalle 2016; Tesfaye et al. 2020); and these purposes are often gendered. Cultural uses can also include the aesthetic value of water (Baker et al. 2015) and opportunities to create aesthetically beautiful spaces such as flower gardens (MacArthur et al. 2022a). Cultural flows or cultural water are water resources and water conditions informed by Indigenous Peoples to sustain and benefit them, improving their spiritual, cultural, environmental, and economic conditions. These uses are likely to be influenced by gender norms specific to the context.
- **Wastewater exposure and reuse.** Water use for different purposes is also associated with wastewater and pollution (Forsey and Bessonova 2020; Kelm et al. 2017), to which there may be differential exposure to associated risks. For example, women's additional contact with polluted water has been shown to create more waterborne infections than men (Adelodun et al. 2021), and women are exposed to skin irritations and infections because of their work producing vegetables and meals (Ungureanu, Vlăduţ, and Voicu 2020). Equally, studies have found boys to have higher exposure to fecal contamination than girls in urban slums because of bathing in contaminated water (Chua et al. 2021). Women and men also have differential opportunity to derive potential benefit where wastewater or its contents (for example, nutrients) are recycled for additional purposes, with calls to increase a focus on women and gender roles (Drechsel, Qadir, and Galibourg 2022; Taron, Drechsel, and Gebrezgabher 2021).

Unsafe water. A woman fills a water jug at a compromised water source (photo by Evgeny - stock.adobe.com).

Socially Differentiated Well-Being Impacts

Water security also impacts the ability of individuals to flourish—what is known in the capability approach as "achievements." These impacts are socially differentiated, with women and vulnerable groups often facing disproportionate negative health impacts, exacerbated in water-related disasters such as drought and flood. This includes (but is not limited to) impacts on physical, psychological, relational, and economic well-being that arise from differentiated experiences with respect to domestic, productive, cultural water availability, accessibility, and use.

- **Physical.** Impacts on physical well-being in relation to nutrition, water quality, and water carrying are strongly socially differentiated with women who are disproportionately affected by water insecurities. Food security and nutrition are directly related to water security through irrigation and water for cooking (Stevenson et al. 2016; Young et al. 2021a), with women and girls often the last to eat, leaving them more at risk of malnutrition (Hathi et al. 2021). Water quality issues also disproportionately impact women. Women have higher rates of cancer, anemia, fertility issues, and skin diseases than men when exposed to poor quality water (Canipari, De Santis, and Cecconi 2020; De Guzman et al. 2023). Water carriers, primarily women, are also susceptible to water-carrying injuries to the neck, spine, and knees, as well as to uterine prolapse and miscarriage (Dickin and Caretta 2022; Hanrahan and Mercer 2019; Kayser et al. 2019; Meierhofer et al. 2022).
- **Psychological.** Impacts on well-being related to mental well-being and stress (Bisung and Elliott 2017). This stress can include feelings of "shame, fear, anger, frustration, embarrassment, feelings of negative identity and marginalization" related to water quality, quantity, or stability (Nunbogu and Elliott 2022, 6). Water-related stresses, such as scarcity for small-scale farming, livelihoods, and fishing, have significant impacts on women who may be dependent on these activities for food, feeding of their families, and livelihoods. Socially differentiated, stresses can also be related to displacement impacts from the construction of hydropower systems in which women are disproportionately impacted by forced migration (Earle and Bazilli 2013).
- **Relational.** Impacts on well-being related to interactions with other individuals, including conflict and violence. This can include intrapersonal conflict and backlash within the home or community, which can escalate to physical violence (Bryan and Mekonnen 2023; Caruso et al. 2021; Nunbogu and Elliott 2022). Importantly, potential negative relational impacts are more potent in water-related projects aiming to empower women (Bryan and Mekonnen 2023).
- **Economic.** Impacts on well-being related to economic and financial security. This can include positive impacts related to income derived from water-connected activities such as irrigation or livestock rearing, and it can also include negative impacts during seasons of drought or water insecurity (van Hoeve and van Koppen 2005; Elliott et al. 2019). An often-overlooked economic aspect is related to education and the opportunity costs that students (primarily girls) face during menstruation and times of water insecurity when water collection is more time-consuming (Agol and Harvey 2018; Swaraj and Maheshwari 2022).

Water-Related Disasters

- **Resources for disaster preparedness, response, and recovery.** Many authors point out the gendered and social differences in access to resources to prepare for water-related disasters. In the face of climate change, various authors point out how existing gendered dependence on natural resources and gender division of labor can result in worsening burdens for women and marginalization of certain groups in crises (Sultana 2009; Myrttinen 2017). In particular, in rural areas, households headed by women have been identified as more vulnerable to disasters (Alhassan, Kuwornu, and Osei-Asare 2019). Other authors describe how early warning systems lack a gender lens and engagement of women in their design and decision-making, meaning that women may be the last to receive or understand information (Tandon et al. 2022). Differential access to skills—such as the ability to swim (Jagnoor et al. 2019)—as well as to transport and communication and to health facilities (Moayed et al. 2020) also create gendered vulnerabilities with respect to disasters. Authors call for a better understanding of differential impact, which needs to be underpinned by gender and age inequality-informed data (Brown et al. 2019).
- **Agency in disaster preparedness, response, and recovery.** Literature points to gender differences in capacity to seek help and support during disasters, such as (i) inadequate authority and confidence to make decisions during crises (Jagnoor et al. 2019) and (ii) limited mobility because of caregiving roles (Moayed et al. 2020). Equally, other authors point to promising developments where women are increasingly recognized for valued roles in flood risk reduction and response, for active roles in early warning systems, and to support water management during scarcity (Tandon et al. 2022; Gero, Chowdhury, and Winterford 2022). A growing body of literature asserts the importance of tapping women's knowledge and capabilities for disaster preparedness and response. For example, studies highlight that women's income sources should be diversified especially during post-disaster periods; yet women's activity remains confined within or close by their village and is conditional on fulfilling their expected roles such as caregiving or household work (Howard 2023).
- **Well-being impacts of disasters.** The death toll among women caused by typhoons, cyclones, and tsunamis is significantly higher than men (Brown et al 2019). There are extensive reports of risks to women's well-being as a result of disasters, particularly with respect to violence and safety (Jagnoor et al. 2019; Moayed et al. 2020). Studies have also shown that women may be more susceptible than men to psychological trauma from disasters (Jagnoor et al. 2019; Moayed et al. 2020). The extreme drought caused by El Niño in 2015–2016 created significant challenges for the people of Vanuatu in accessing safe drinking water, especially affecting women and children. Community consultation highlighted that women carry the largest burden during a drought, as they are responsible for collecting water for themselves and their children, and for household duties. Unlike men, who could bathe publicly during the 2015–2016 drought, women were unable to maintain basic hygiene because of cultural and safety concerns (Grant, Willetts, and Huggett 2019).

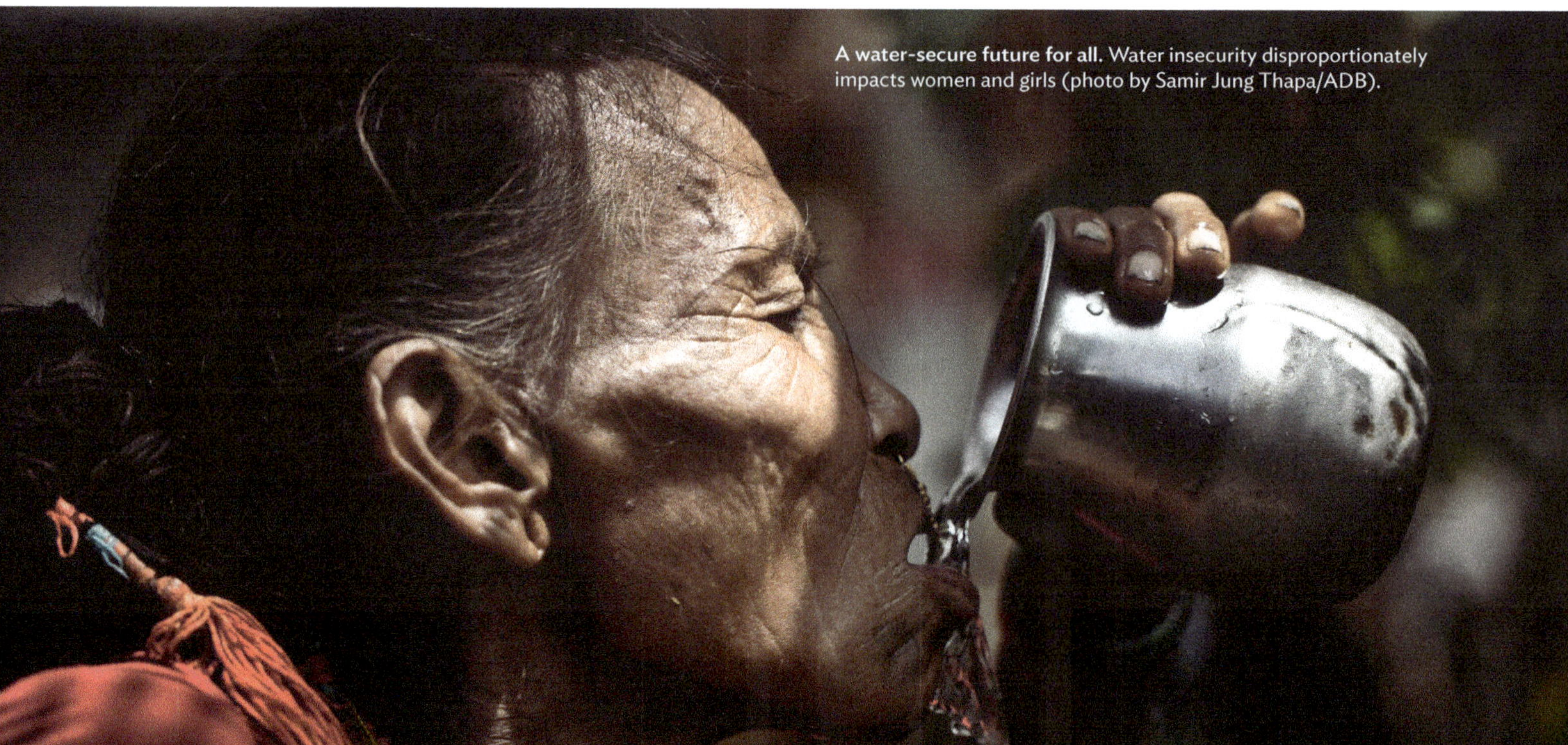

A water-secure future for all. Water insecurity disproportionately impacts women and girls (photo by Samir Jung Thapa/ADB).

IV. Toward Transformational Practices

This section presents case studies that can inform the design of initiatives supporting women's inclusion in countries where women's water security and gender equality require attention. Integration of gender into water security can be done up to different extents (Figure 4) (MacArthur et al. 2023). This gender integration framework overlays four categories of gender integration with associated motivations and tendencies. The framework demonstrates different available entry points for design or implementation teams and related stakeholders to discuss during program design. Depending on the level of stakeholder buy-in, skills, and resources to address gender integration, approaches can be designed to be gender sensitive, gender responsive, or gender transformative.

- **As a minimum, moving from gender-insensitive to gender-sensitive approaches.** When the main focus of program outcomes is on technical aspects of water security rather than on social or gender outcomes, gendered aspects are often approached with an "instrumental" tendency aligned to motivations of welfare and efficiency. When this is the case, as a minimum, it should be a key priority to shift from gender-insensitive to gender-sensitive approaches that acknowledge and work within existing gendered norms, dynamics, and structures with sensitivity and awareness.
- **Taking up opportunities for transformative change.** For programs interested in leveraging the transformative potential of water security programs to achieve wider gender and social outcomes, and where high-level commitment, budget, and expertise are available, gender-responsive and gender-transformative approaches can be adopted. These approaches are motivated by a focus on equity, empowerment, and equality, and they seek to question, address, and shift existing gender norms, dynamics, and structures.

Figure 4: A Framework Toward Gender-Transformative Practice

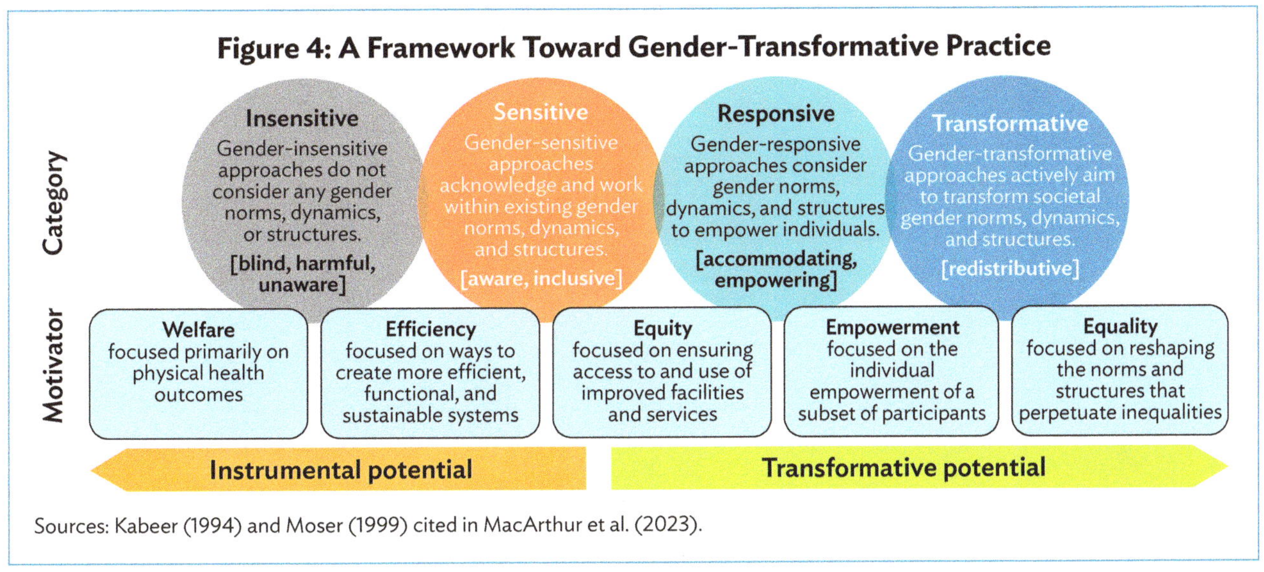

Sources: Kabeer (1994) and Moser (1999) cited in MacArthur et al. (2023).

All efforts to address gender inequality can result in resistance and backlash; hence, gender-sensitive, gender-responsive, and gender-transformative approaches should be accompanied by robust "do no harm" strategies (Boxes 2 and 3).

In subsections 4.1–4.4, 33 example case studies of gender integration are provided and categorized as gender sensitive, responsive, or transformative. The case studies include information about the adopted strategy, relevant outputs, and outcomes. These case studies were identified through a literature exploration process and in consultation with the Inclusion Roundtable that was launched at Stockholm World Water Week in 2022 hosted by ADB. Priority was given to published and publicly available case studies, with a focus on identifying a breadth of relevant strategies. This list is not intended to be exhaustive but illustrative of the different modalities for integrating gender into water security programs.

The case study examples cover four key areas relevant to water security: security: (i) enhancing women's water and sanitation access (subsection 4.2; Tables 1 and 2), (ii) reducing women's vulnerability to water-related disasters (subsection 4.3; Table 3), (iii) improving equitable water management and access to irrigation water (subsection 4.4; Tables 4 and 5), and (iv) strengthening women's employment in water organizations (subsection 4.5; Tables 6 and 7).

Summary of Gender Integration Strategies

The 33 cases represent programming that adopts one or more of the five broad strategies. Many of the cases align with multiple strategies as a best practice.

Embed gender thinking in policy considerations. This reflects the significance of gender within issues of water security and includes work to strengthen policies for inclusive infrastructure design (case 3), urban master plans (case 6), disaster early warning systems (cases 13 and 14), roles in water use committees and clubs (cases 18 and 21), and utility policies (cases 24 and 25).

Box 2: Do No Harm

Regardless of the category or motivation, all gender-integrated approaches are at risk of causing backlash and potential harm to participants and communities. As such, it is important to adopt a robust "do no harm" strategy that purposefully interrogates potential unintended harm and incorporates mitigation measures. In some cases, poorly implemented transformative approaches may be more harmful than well-considered sensitive approaches, as it is not always appropriate to engage transformative change unless the relevant skills and resources are available. Transformative approaches require appropriate staff training and more complex forms of monitoring and evaluation.

Source: ADB.

Box 3: Gender-Transformative Approaches

The water sector is increasingly adopting approaches toward the "transformative" end of the gender-integration framework (Figure 3). Transformative water security practices require focus on both water security and gender equality objectives. A gender-transformative approach is often characterized by five tenets (MacArthur et al. 2022b):

- **Why**—starting from a purposefully gender-transformative agenda toward social change
- **Where**—engaging with structural systems and not just the symptoms of inequalities
- **What**—maintaining a focus on strategic gender interests beyond practical gender needs
- **Who**—recognizing and valuing diverse identities through an intersectional perspective
- **How**—adopting participatory, collaborative, and intentional methodological approaches

Source: ADB.

Promote meaningful engagement, promote codesign and participation, and address barriers. Promote meaningful engagement of women, girls, and other marginalized individuals, and avoid harmful token participation. This can include the participatory design of infrastructure, products, standards, programs, and services (cases 1, 2, 3, 4, and 23), participatory implementation (cases 5, 7, 9, 12, 15, 16, 17, 18, 19, 20, 21, and 24), and participatory monitoring and evaluation (cases 8 and 13).

Increase women's voice, leadership, and decision-making in water and sanitation service provision and wider water management. Doing so empowers women working and volunteering in water and sanitation service provision through support of their leadership, voice, and decision-making. This includes activities in the private sector (cases 22, 31, 32, and 33), the public sector (case 25), civil society (case 28), and communities (cases 5, 16, 17, and 18).

Address systemic inequality within organizations. This can be done within formal and informal water-related organizations through analysis (cases 29 and 30), dialogue and training such as unconscious bias (cases 14 and 26), strategic partnerships (case 10), and celebration of champions (cases 27 and 28).

Monitor and evaluate gendered outcomes and impacts, both intended and unintended. These actions pertain to impacts on programs, policies, activities, and institutions. These processes include and go beyond sex-disaggregated data and include intersectional aspects, practical gender needs, and women's strategic gender interests. This requires a foundation of strong gender analysis (cases 4 and 6) to look at individual impacts (cases 8, 11, and 13) and institutional trends (case 29), as well as pay attention to unintended outcomes.

Enhancing Women's Access to Water Supply and Sanitation

As women and girls are traditionally responsible for household water and sanitation tasks, there are significant opportunities to partner water supply and sanitation interventions with gender equality and inclusion initiatives. Such incentives address water security aspects like water availability, access, and use, while being cognizant of the societal and institutional contexts. Examples are provided about facility and infrastructure design and the integration of gendered thinking into community water, sanitation, and hygiene (WASH) activities.

Embed gender considerations in facility and infrastructure design and implementation

There is potential to meaningfully include women and marginalized community members in the design of WASH facilities and infrastructure through human-centered and codesign processes. The examples below are all gender sensitive, as they do not aim to explicitly empower women nor transform gender norms and structures. However, women's involvement in facility and infrastructure design can enable gender transformation. For instance, the first case in Table 1 also included engagement with governments across multiple countries, which has the potential to shift broader norms in facility design.

Enhancing women's access to water. Women are often responsible for water collection and management within households (photo by Al Benavente/ADB).

Table 1: Enhancing Women's Access to Water Supply and Sanitation—Embed Gender Considerations in Facility and Infrastructure Design and Implementation

No.	Objective	Key Areas	Category
1	Promote codesign of water, sanitation, and hygiene (WASH) infrastructure with and for women—Fiji and Indonesia (Leder et al. 2021)	• **Strategy.** The Revitalising Informal Settlements and their Environments (RISE) program codesigned drainage and sanitation systems in informal settlements with community members taking an intersectional gender perspective. • **Outcomes.** The tool kit was piloted in 24 informal settlements across Indonesia and Fiji with adjustments made to designs based on diverse inputs. The tool kit includes a policy brief to support other countries in adopting codesign processes. • **Resources.** A tool kit in English and Bahasa Indonesia and an example process of including diverse individuals in the development of complex sanitation infrastructure.	**Gender sensitive** Codesigning community infrastructure aimed to meaningfully include the voice of women and other marginalized community members.
2	Leverage human-centered design (HCD) strategies to design inclusive facilities—Bangladesh, Cambodia, and Zimbabwe hcdforwash.org	• **Strategy.** In Bangladesh, Cambodia, and Zimbabwe, PRO-WASH and iDE have supported the integration of gender-sensitive HCD to support handwashing behavior change strategies, tube-well platform designs, and latrine design. • **Outcomes.** For example, 10,000 innovative low-cost tube-well platforms were sold in Bangladesh during 2014–2019. • **Resources.** The HCD for WASH online tool kit provides additional case examples, tools, and templates to promote HCD activities within the WASH sector.	**Gender sensitive** Gender-sensitive HCD aims to meaningfully include women and other marginalized community members in the design of a variety of WASH products and strategies.
3	Normalize female-friendly toilets through standards and guidance—Bangladesh, India, Nepal, and Tanzania (UNICEF, WaterAid, and Water and Sanitation for the Urban Poor 2018)	• **Strategy.** The United Nations Children's Fund (UNICEF), WaterAid, and Water and Sanitation for the Urban Poor created a guide to support planners and decision-makers in the design and operation and maintenance of female-friendly public and community toilets. The guide was collaboratively developed with women in four countries. • **Outcomes.** The guide outlines essential and desirable features that make toilets female friendly and has been used in four countries to encourage policy change. • **Resources.** Guidance around the design of female-friendly community latrines in English, French, and Spanish and a replicable process of actively including women in the design of female-friendly infrastructure.	**Gender sensitive** This toilet guidance development process aimed to meaningfully include women to meet their practical toilet needs.
4	Utilize tools and guidelines for analysis, monitoring, and evaluation of gender mainstreaming in infrastructure—Burkina Faso, India, Mozambique, Senegal, and Tanzania	• **Strategy and resource.** The Japan International Cooperation Agency's Reference Material for Gender Mainstreaming in the Water Resources Sector provides guidance for Japan International Cooperation Agency's grant aid, official development assistance loans (soft loans), and technical cooperation projects (all infrastructure and capacity development). The guideline outlines aspects such as social and gender analysis, action plans, establishment of indicators, gender monitoring, and gender evaluation. Similar tools have also been created by the World Water Assessment Program for gender analyses and the collection of sex-disaggregated water data. • **Outcomes.** The guideline has been used in supporting the development of rural water supply facilities in African and South Asian countries such as Senegal, Mozambique, India, Burkina Faso, and Tanzania. Specific case examples are included in the guideline.	**Gender sensitive, gender responsive** Gender analysis processes enable teams to foster inclusivity and empowerment.
5	Foster inclusive and gender-focused, and sustainable water and sanitation service delivery—Nepal	• **Strategy.** In Nepal, the Asian Development Bank (ADB) has helped provide quality water and sanitation services in small towns. The project supported the construction and upgrades of intake, reservoirs, water pipes, treatment facilities, and household connections, as well as septic tanks, drainage, and public awareness campaigns. The process aimed to meaningfully involve women and other marginalized groups in planning, implementation, and evaluation. • **Outcomes.** The project had practical and strategic gender equality benefits. Practical benefits included improvements in access to water supply and sanitation infrastructure. Strategic benefits included women's economic empowerment, and more equal decision-making, leadership, and leisure time. • **Resources.** The project has produced case studies on the gender equality results as well as the service delivery model.	**Gender responsive** The meaningful engagement of women and vulnerable groups led to empowerment impacts and other strategic outcomes.

continued on next page

Table 1 *continued*

No.	Objective	Key Areas	Category
6	Support improved municipal service delivery and urban governance in project towns through gender action planning—Bangladesh, Georgia, Cambodia, and Tajikistan	• **Strategy.** To support better integration of gender into municipal water infrastructure, gender action plans have been developed to help plan, implement, and monitor gender outcomes. The plans have been used to support water infrastructure in Bangladesh, Cambodia, and Tajikistan. • **Outcomes.** For example, in Bangladesh, the process has led to gender-inclusive urban master plans in 17 towns, improved capacity of local governments to support urban service delivery, and the development of gender- and climate-responsive municipal infrastructure. • **Resources.** This ADB series of five quick guides is designed to support staff and consultants in implementing gender action plans. An example action plan for Bangladesh provides details of an action plan for water infrastructure.	**Gender responsive** Gender action plans identify opportunities to increase women's meaningful participation in project activities, including planning, implementation, and monitoring of infrastructure.

Source: ADB.

Support gender-integrated community activities and expanded monitoring

Community water and sanitation activities are a common strategy to influence behavior and foster lasting WASH change. As highlighted in cases 7 to 9, community water and sanitation activities can be adapted to promote gender-transformative change by actively aiming to address gender inequalities in household or community discussions.

Table 2: Enhancing Women's Access to Water Supply and Sanitation—Support Gender-Integrated Community Activities and Expanded Monitoring

No.	Objective	Key Areas	Category
7	Encourage participatory activities within dialogue groups to transform gender roles—Fiji and Vanuatu (Halcrow et al. 2010)	• **Strategy.** In Fiji and Vanuatu, Live and Learn and World Vision utilized a set of participatory activities to encourage community dialogue around gender inequalities. These were documented in a resource guide for future use. • **Outcomes.** The use of the activities encouraged communities to explore decision-making and participation, and to value gender differences within and beyond water and sanitation management. • **Resources.** The Working Effectively with Women and Men tool kit includes a resource guide, poster, flash cards, and detailed case examples from Fiji and Vanuatu.	**Gender transformative** Participatory community water, sanitation, and hygiene (WASH) activities actively aimed to transform gender norms and dynamics.
8	Promote transformative monitoring of gender dynamics to raise awareness and monitor outcomes—Viet Nam (Leahy et al. 2017; Lee et al. 2018)	• **Strategy.** In Viet Nam, Plan International developed a tool to raise awareness on gender roles and responsibilities and evaluate change as part of sanitation programming. • **Outcomes.** Reported benefits included changes in household workloads, increased confidence for women, improvements in communication, and more women occupying leadership roles. • **Resources.** The Gender and WASH Monitoring Tool can be used to spark community discussions about gender inequalities and to monitor change.	**Gender transformative** This gender-focused monitoring tool actively aimed to transform gender norms and dynamics.
9	Facilitate gender-focused discussions in community-led total sanitation (CLTS)-triggering activities—Timor-Leste (Grant and Megaw 2019)	• **Strategy.** In Timor-Leste, WaterAid facilitated gender-focused discussions to provide opportunities for community and household dialogue on gender inequalities alongside water management and CLTS activities. • **Outcomes.** Undertaken in 56 locations, the dialogues led to increased valuing of gendered workloads, increased status of women in communities, and renegotiated water- and sanitation-related roles. University of Technology Sydney's Institute for Sustainable Futures has undertaken an evaluation of the approach. • **Resources.** WaterAid has produced a discussion manual on how to integrate gender into WASH activities and lead gender discussions in communities.	**Gender transformative** These adapted CLTS and water management activities actively aimed to transform gender norms and dynamics.

continued on next page

Table 2 *continued*

No.	Objective	Key Areas	Category
10	Partner with rights holder organizations—Asia and the Pacific region (Water for Women 2022b)	• **Strategy.** The Water for Women Fund has promoted the value of partnerships between WASH organizations and rights holder organizations. Rights holder organizations, which are made up of advocates for marginalized groups, are key to ensuring that this happens. They include organizations that support the rights of diverse women, people with disabilities, sexual and gender minorities, ethnic minorities, or the economically disadvantaged. • **Outcomes.** Increased empowerment and involvement of marginalized people, and strengthening of the organizations that represent these communities, leading to more inclusive WASH outcomes. • **Resources.** This Water for Women guide provides recommendations to strengthen interorganizational partnerships.	**Gender transformative** The Water for Women Fund's partnership approach aimed to transform gender norms within organizations.
11	Monitor gender-differentiated impacts on water security—Global (Carrard et al. 2022; Jepson et al. 2017b; JMP 2023; Sinharoy et al. 2023; Young et al. 2021b; Young et al. 2019)	• **Strategy.** The Joint Monitoring Programme of UNICEF and the World Health Organization highlights the importance and challenges of monitoring gender and socially differentiated impacts on water security, sanitation, and hygiene at the household level. The program provides standardized indicators suitable for use for global reporting and suggests the use of tools and strategies beyond sex-disaggregated data at a global scale. Examples of such tools include the Household Water Insecurity Experiences (HWISE) and Individual Water Insecurity Experiences (IWISE) scales (Young et al. 2019; Young et al. 2021b); the WASH-Gender Equality Measure (GEM) (Carrard et al. 2022); and the Agency, Resources, and Institutional Structures for Sanitation-Related Empowerment (ARISE) scales (Sinharoy et al. 2023), which offer opportunities to measure gender-related impacts on water and sanitation programming. • **Resources.** The Joint Monitoring Programme conducted a review of gender-related measures, providing an inventory and a report. The household and individual water security experiences measures are covered in HWISE/IWISE resources. The WASH-GEM resources and training materials support the implementation of this tool at the local level. The ARISE scales are described here.	**Gender sensitive, gender responsive, gender transformative** Tools to monitor gender-differentiated impacts range from sensitive to transformative, focusing on aspects of inclusion, empowerment, and norms change.

Source: ADB.

Reducing Women's Vulnerability to Water-Related Disasters

Women, girls, and marginalized groups are often disproportionately impacted by water-related disasters and often lack resources and agency; they therefore experience fewer positive outcomes. Four community resilience examples are provided, which highlight the value of participatory activities in preparing for disasters. The examples include both gender-responsive and gender-transformative initiatives, with gender-transformative initiatives partnering resilience and gender programming.

Improving Equitable Water Management and Access to Irrigation Water

There is significant evidence that improvements in economic security for women and marginalized groups are connected to improved agricultural productivity and irrigation in rural communities. However, irrigation water is often mechanized and managed by men. As such, there are significant opportunities to increase women's meaningful involvement in the management and design of irrigation and multiple use systems.

Support community-led management for irrigation

As explored in these cases, the meaningful involvement of women in irrigation activities can be implemented at macro, meso, and micro levels. The extent to which the activities were gender transformative has been directly related to the program theories of change and the degree to which societal change was pursued.

Table 3: Reducing Women's Vulnerability to Water-Related Disasters

No.	Objective	Key Areas	Category
12	Utilize participatory activities to address complex vulnerabilities—Kenya, Peru, and India (Nyukuri, Naess, and Schipper 2016; Arana, Quezada, and Clements 2016; Sogani 2016)	• **Strategy.** In Kenya, India, and Peru, participatory activities aimed to address the complexity of climate risks and vulnerability. This included the strengthening of self-help groups to understand vulnerability and participatory planning processes. Innovation in monitoring and evaluation ensured appropriate tools, and progress indicators included aspects such as gender-based violence, gendered unemployment, accessibility to services and opportunities for both genders, and skills transfer. • **Outcomes.** Improvements in well-being, productivity, and living conditions for diverse populations and increased participation of women in program activities and planning processes. In India, activities increased confidence between community members and volunteers to address weak social cohesion, which exacerbates gender inequalities. • **Resources.** Building on these experiences, the Climate and Development Knowledge Network's Advancing Gender Equality and Climate Action practical guide supports programs integrating gender into climate change programming. The guide is supplemented with a gender training resource pack for practitioners and program teams.	**Gender responsive** Participatory activities aimed to empower women and other marginalized community members to mitigate climate risks.
13	Foster gender-transformative early warning systems—Philippines, Peru, and Nepal (Brown et al. 2019; AASCTF 2021)	• **Strategy.** Programs in Nepal, Peru, and the Philippines adopted a gender-transformative approach to early flood warning. The projects applied a participatory story collection methodology to support the design of a smart flood early warning system alongside the goal of transforming social norms such as decision-making and participation. • **Outcomes.** The process led to the development of a multipronged early warning system including redesigned shelters and improved testing of materials. • **Resources.** The Missing Voices story collection methodology can help project teams listen to marginalized gender groups and more effectively respond to intersectional needs.	**Gender transformative** The innovative process led to the design of a flood warning system that also aimed to transform social norms.
14	Encourage raised critical consciousness and consensus from resilience consortium leadership—Ethiopia and Burkina Faso (McOmber, Audia, and Crowley 2019)	• **Strategy.** The Building Resilience and Adaptation to Climate Extremes and Disasters (BRACED) program conducted a series of workshops with consortium partners to build critical consciousness of program teams and to create a shared understanding of a gender-transformative perspective of resilience. This foundation is critical for changing gender norms and transforming organizations. • **Outcomes.** The workshops led to harmonized consortium perspectives around the importance of gender-transformative approaches and development of targeted activities. • **Resources.** While not used in the BRACED program, the Water for Women Fund's Gender Equality and Social Inclusion Self-Assessment Tool can be used by organizations to internally assess critical consciousness.	**Gender transformative** The consortium focused on transforming gender norms and structures within the implementing organizations.
15	Form gender-transformative early warning committees—Ethiopia and Burkina Faso (McOmber, Audia, and Crowley 2019)	• **Strategy.** In Burkina Faso and Ethiopia, a consortium of organizations through the BRACED program aimed to address inequalities between men and women alongside and through resilience-building initiatives. This included supporting women's income generation and establishing early warning committees that meaningfully engaged women. • **Outcomes.** The early warning committees increased women's capacity, visibility, and involvement in village-level activities.	**Gender transformative** The process of establishing early warning communities was conducted to both empower women (gender responsive) and transform the role of women in society.

Source: ADB.

Table 4: Improving Equitable Water Management and Access to Irrigation Water-Support Community-Led Management for Irrigation

No.	Objective	Key Areas	Category
16	Engage women in managing aquifer recharge projects—India	• **Strategy.** In Gujarat and Rajasthan, the Managing Aquifer Recharge and Sustaining Groundwater Use through Village-level Intervention (MARVI) project, led by Western Sydney University, is engaging and empowering communities in India to monitor, use, and manage groundwater. Community training has helped fill knowledge gaps and empower decision-making at the village level. • **Outcomes.** The MARVI project has been successful in ensuring the active participation of women, highlighting women's voices as critical to this success. • **Resources.** The MARVI website includes strategy documentation; a resource library; and community stories in English, Hindi, and Gujarati.	**Gender responsive** The MARVI project was effective in amplifying the voices and participation of women within the community management structures.
17	Promote women's involvement in community-led watershed management—India (Padmaja, Leder, and Garg 2018; Swaraj and Maheshwari 2022)	• **Strategy:** In Central India, the community watershed project of the International Crops Research Institute for the Semi-Arid Tropics and the Central Research Institute for Agro-Forestry aimed to increase drought resilience of farming through groundwater recharge and agroforestry interventions. The community-led watershed management approach actively aimed to include women and marginalized individuals in the management process. • **Outcomes.** The process had positive impacts of women's empowerment and productivity and anticipates impacts on girls' school absenteeism.	**Gender responsive** The community-led management processes aimed to promote women's empowerment and meaningful involvement.
18	Support women in water user committees—Peru (De Nys et al. 2013)	• **Strategy.** In Peru, the Ministry of Agriculture aimed to strengthen the role of women in water user committees through a participatory gender diagnostic, dialogues, and collaborative target setting. The gender diagnostic operated as a gender analysis to explore the gender norms, challenges, and opportunities for irrigation. • **Outcomes.** The gender diagnostic and subsequent refinement of the water committee led to improvement in women's technical skills, self-esteem, and positions in the water users' organizations. The process also "raised awareness among the community members about women's specific needs and expectations related to water management for irrigated agriculture."	**Gender responsive** The Government of Peru sought to empower women as active participants in water user committees.
19	Leverage participatory tools to strengthen partnerships and community trust—Zambia (Poutiainen and Mills 2014)	• **Strategy.** In Zambia, the Irrigation Development Support Programme promoted partnerships among communities, the private sector, and the government for large-, medium-, and small-scale irrigation using participatory activities such as checklist questions to identify constraints to women's participation. The activities encouraged diverse and transparent participation from communities and enabled women to articulate challenges and opportunities to male relatives and project staff. • **Outcomes.** Sensitization of communities on gender issues in irrigation led to increased participation of women as well as policy recommendations regarding affirmative action and land ownership. Another outcome was increased recognition from irrigation leadership of the value of women's involvement in the sector. • **Resources.** In English and Spanish, the Gender in Agriculture Sourcebook provides activities, case studies, and frameworks that the Zambia team drew upon to strengthen partnerships.	**Gender transformative** The collaborative partnership activities aimed to transform gender norms within communities and within participating organizations.
20	Strengthen community leadership for small-scale irrigation—Ethiopia (CARE International 2022)	• **Strategy.** In Ethiopia, CARE International's Water for Food Security, Women's Empowerment and Environmental Protection (SWEEP) project supported women's community leadership in small-scale irrigation alongside action committees, infrastructure development, savings groups, and collaboration platforms. • **Outcomes.** The project saw an increase in household incomes by 8.5 times, higher levels of food security, and time savings for women. The project also increased participation of women in major household financial decisions, increased acceptance of women's leadership in communities, and increased male involvement in domestic chores. • **Resources.** CARE International's Social Analysis and Action Global Implementation Manual provides practical guidance for programs adapting action committees to address gendered community challenges (e.g., irrigation).	**Gender transformative** Leveraging social analysis and action processes aimed to transform gender norms within communities and empower women and other marginalized groups as leaders in small-scale irrigation.

continued on next page

Table 4 *continued*

No.	Objective	Key Areas	Category
21	Promote informal groups to foster community discussions around irrigation management—Sub-Saharan Africa (FAO 2019)	• **Strategy.** Supported by the Food and Agriculture Organization of the United Nations (FAO), Dimitra Clubs are voluntary informal groups for women, men, and youth that discuss and advance agriculture, climate change, education, health, infrastructure, nutrition, peace, and women's rights. • **Outcomes.** Impact has been assessed in many areas, including but not limited to food security and nutrition, gender equality and women's leadership, resilience, peace, and climate change adaptation. • **Resources.** This collection of reports, cases, and webinars provides examples and evidence of the value of Dimitra Clubs.	**Gender transformative** Dimitra Clubs have successfully transformed gender norms within communities toward equality.

Source: ADB.

Table 5: Improving Equitable Water Management and Access to Irrigation Water Promote Multiple-Use Water Systems

No.	Objective	Key Areas	Category
22	Encourage uptake of multiple-use water systems for enterprise development—Senegal (van Houweling et al. 2012)	• **Strategy.** In Senegal, the World Bank utilized community activities to encourage the development and uptake of multiple-use water systems with an explicit goal of supporting women-led water-based enterprises. • **Outcomes.** Multiple-use systems aided in expanding and diversifying women's livelihoods through time savings and greater quantities of water. Notably, women strengthened existing enterprises and started new ones.	**Gender responsive** Multiple-use water systems aimed to strengthen women's economic empowerment through enterprise development.
23	Promote codesign of multiple-use water systems to support women's decision-making—Nepal (van Koppen et al. 2021)	• **Strategy.** In Nepal, the International Water Management Institute engaged diverse community members, including women, in the codesign of multiple-use water systems and management strategies. Similar strategies have also been used by iDE in Nepal. • **Outcomes.** Participatory processes that actively engage women have led to increased women's involvement in decision-making and codesign of systems. This has led to more women-friendly technical designs and better collaboration regarding household income and water-related activities. • **Resources.** The iDE Guidelines for Planning, Design, Construction and Operation of Multiple Use Water Systems (MUS) builds on more than a decade of multiple-use water system activities in Nepal.	**Gender responsive** Codesign processes aimed to include women in the design of multiple-use systems (gender sensitive) and management structures. This approach explicitly aimed to empower women as decision-makers in the process.

Source: ADB.

Promote multiple-use water systems

Multiple-use water systems offer opportunities to address both domestic and economic challenges in water access and use. The two examples provided are gender responsive as they aim to empower diverse voices in the processes of design, uptake, and management of water systems. However, programs could become gender transformative if they actively aim to address gender norms, dynamics, or structures in the community more broadly.

Strengthening Gender Equality in Water Organizations

This last category of case studies explores the formal and informal employment of women in water and sanitation organizations harmonizing with wider objectives of increasing women's representation in government and the private sector. Employment in formal institutions is directly shaped by institutional and societal contexts, while informal employment such as entrepreneurship is shaped by water security more broadly alongside societal influences.

Encourage participation in formal institutions and the workforce

Changing harmful norms related to women's involvement in the water-related workforce requires attention and action at every stage of the career cycle and starts in childhood. Examples of responsive and transformative practice include encouraging girls to study subjects related to roles in the water sector, deliberately addressing unconscious bias in recruitment, and addressing barriers to career advancement for people who need to take time off for family care roles.

Table 6: Strengthening Gender Equality in Water Organizations—Encourage Participation in Formal Institutions and the Workforce

No.	Objective	Key Areas	Category
24	Community-based water management in urban poor settlements—Manila, Philippines	• **Strategy.** The Pag-asa sa Patubig (Hope in Water) Partnerships water-for-the-poor program adopts a community-based approach in the management of bulk water systems in urban poor settlements. Community members, mostly women, are trained to become officers of cooperative-based water associations that run the day-to-day operation of these community systems. • **Outcomes.** Apart from overcoming financial, legal, and technical obstacles in bringing affordable potable water to urban poor settlements, women members of these communities now enjoy improved hygiene conditions and increased productivity with the convenience of better water access. The engagement also provides women the opportunity to establish themselves as community leaders and mobilizers.	**Gender sensitive** Actively involving women in community-based water projects allows them to heavily influence decisions and policies about water access and related needs such as hygiene management.
25	Water-focused social entrepreneurship—Manila, Philippines MWSS (2023)	• **Strategy.** Complementing the bulk water system that Maynilad established in the urban poor settlement of Riverview, the company engaged the women of the community for the Green Badge Uniform Upcycling Program. This augmented the income of the women, mostly housewives, and allowed them to continue to afford the basic needs of the household, including water. • **Outcomes.** The pilot implementation of the Green Badge program in 2017 allowed the women to earn ₱2,000–₱4,000 ($36–$72 as of 2017) per transaction with Maynilad. This gave the housewives more financial independence from their husbands, as they were able to spend the money on water, additional food, the schooling needs of their children, and modest personal leisure.	**Gender responsive** Combining a water access project with a livelihood program enhances the sustainability of both interventions, which have significantly improved the welfare and agency of women in these communities.
26	Provide unconscious bias training to water institutions to address systemic inequalities—Melbourne, Australia (Melbourne Water 2022)	• **Strategy.** Melbourne Water provides education and training to all managers on unconscious bias, its impacts at work, and how they can amend their behavior to manage biases. • **Outcomes.** Employees and leaders increase their understanding of unconscious bias, its impacts at work, and how they can amend their behavior to manage biases. • **Resources.** Melbourne Water has drawn on the University of Technology Sydney's Institute for Sustainable Futures (UTS-ISF) inclusive workforce guidance and its own inclusivity strategy. UN Women also provides online self-paced courses to aid teams in addressing knowledge and bias barriers.	**Gender responsive** Bias training proactively aims to address barriers within water organizations, creating more inclusive workplaces.
27	Celebrating diverse leadership in the water, sanitation, and hygiene (WASH) workforce—Global (IWA 2022)	• **Strategy.** The International Water Association Gender Diversity and Water Award acknowledges and celebrates female leadership by recognizing women working in the field of water who, through their demonstrated leadership, have had a significant positive impact on the development of the industry. • **Outcomes.** Achievements include promoting positive role models and measuring their impact on gender diversity advancement (e.g. number of people, scale of transformation) in connection with relevant policies, programs, and other activities. • **Resources.** The International Water Association provides case studies on each year's winner.	**Gender responsive** Celebration of diverse and empowered leadership creates role models and pathways for future leaders.

continued on next page

Table 6 *continued*

No.	Objective	Key Areas	Category
28	Support transformative leadership for inclusive WASH—Bhutan (Gonzalez et al. 2022)	• **Strategy.** In Bhutan, SNV aimed to build potential for transformative leadership in the Ministry of Health's Rural Sanitation and Hygiene Programme. The project invested in women's leadership skills, established networks, and worked with men to increase women's voice and influence in the WASH sector. • **Outcomes.** The project saw personal outcomes for individual leaders, increased critical consciousness, and examples of leaders taking transformative action to support equality and inclusion. • **Resources.** The experience was documented in a journal article. The project also created a series of seven videos to inspire transformative leadership.	**Gender transformative** Transformative leadership in water and sanitation demonstrates how leaders can influence gender norms, structures, and dynamics.
29	Foster opportunities to deepen gender dialogue and support inclusion in the water sector—Global (World Bank 2019)	• **Strategy.** The World Bank has created the Equal Aqua collaborative platform, which aims to deepen the dialogue on gender diversity and inclusion in water sector jobs by (i) connecting utilities, associations, and representatives from the private sector, academia, and local and international organizations; and (ii) benchmarking gender inclusion in water organizations. • **Outcomes.** The platform has brought together 19 organizations to share knowledge, best practices, and support. • **Resources.** The Equal Aqua report highlights the challenges of gender parity and diversity inclusivity in the sector across the globe.	**Gender sensitive, gender responsive, gender transformative** The collaborative platform creates opportunities for inclusion, dialogue, and systemic change within water organizations.
30	Diagnose and shape gender dynamics within water and sanitation institutions—Global (Kumar, Grant, and Willetts 2021)	• **Strategy.** Guidance and resources have been developed to support inclusive workplaces in the water sector, including diagnosis of gender parity and gender dynamics, and strategies to address attraction, recruitment, retention, and advancement of diverse employees, particularly women and other marginalized groups. • **Resources.** UTS-ISF has compiled a database of inclusive WASH workforce strategies and guidance to strengthen inclusivity in WASH organizations.	**Gender sensitive, gender responsive, gender transformative** Diagnosing and addressing specific inclusion barriers against the whole career cycle aim to actively make organizations more inclusive. Strategies include sensitive, responsive, and transformative approaches.

Source: ADB.

Gender-inclusivity in water projects. As women and girls are traditionally responsible for household water and sanitation tasks, there are significant opportunities to partner water supply and sanitation interventions with gender equality and inclusivity initiatives (photo by Chor Sokunthea/ADB).

Support women in enterprises and entrepreneurship

There is an increasing number of opportunities for small-scale enterprise and entrepreneurship in WASH and research shows low participation of women.

There are strong imperatives to involve women and marginalized peoples as part of an inclusive approach, to ensure the needs of the whole community are addressed, and so that women may access jobs and leadership and economic opportunities offered through water and WASH enterprises.

Table 7: Strengthening Gender Equality in Water Organizations—Support Women in Enterprises and Entrepreneurship

No.	Objective	Key Areas	Category
31	Support women in piped-water enterprises—Cambodia (Grant et al. 2017; Grant et al. 2019)	• **Strategy.** In rural Cambodia, women have established piped-water enterprises, and research has shown that technical and financial support, capacity development, and networking can aid in empowering women entrepreneurs to run successful businesses and transform gender norms within the communities they service with piped water. • **Outcomes.** The Cambodian Water Association has provided training and support for piped-water enterprises. Related research notes that this can (but does not always) support empowering outcomes. • **Resources.** This synthesis of women's water, sanitation, and hygiene (WASH) entrepreneurship in Cambodia includes examples of piped-water enterprises.	**Gender responsive** Targeted training and support for women who deliver piped-water enterprises can enhance empowerment outcomes.
32	Support women in WASH enterprises—Cambodia and Indonesia (Indarti et al. 2019; Soeters et al. 2020; Water for Women 2022a)	• **Strategy.** Similarly, across Southeast Asia, a variety of organizations have been supporting women in starting or expanding small WASH enterprises. An example is a partnership between WASH organizations and SHE Investments, a business incubator that supports female entrepreneurs in Cambodia. SHE Investments provided business training, mentoring, financing, and networking opportunities for women from all business sizes and sectors, including WASH. • **Outcomes.** In Cambodia, women-focused training found that incomes for women and their families were increased and that all business owners were regularly tracking finances after the program. Savings were increased and debts were reduced. Women's confidence and decision-making power improved, with increased networking opportunities created and female leadership encouraged. In Indonesia, attention to women's empowerment and to intersectionality has resulted in more diverse women achieving business success. • **Resource.** This synthesis of efforts of multiple organizations in Cambodia highlights examples of women-focused support.	**Gender responsive** Targeted training and support aim to empower women-led WASH enterprises.
33	Transform wife–husband sanitation businesses—Cambodia (MacArthur and Koh 2022; Soeters et al. 2020; Water for Women 2022a)	• **Strategy.** Building on the previous example, iDE Cambodia, in partnership with SHE Investments, has been supporting women and men (often in husband–wife partnerships) in operating sanitation enterprises. The support involves supporting women and men differently, with an aim of transforming social norms alongside strengthening sanitation businesses. • **Outcomes.** The jointly run Cambodian businesses have seen not only improvements in business sustainability and success, but also improvements in wife–husband collaboration and communication as documented in this "photovoice" study ("photovoice" is a participatory photography approach that encourages participants to take photos through their own "lens"). • **Resources.** This synthesis of efforts of multiple organizations in Cambodia highlights examples of husband–wife support for sanitation businesses.	**Gender transformative** Targeted training and support for wife–husband sanitation businesses aim to not only empower women entrepreneurs, but also to address household-level gender norms, dynamics, and structures.

Source: ADB.

V. Conclusions and Recommendations

The conclusions of this report are threefold:

- A change in mindset and conceptualization of water security from a biophysical or macro-level construct to a social one that foregrounds the human dimensions of water security is fundamental to increased consideration of gender equality and inclusion in water security.
- Efforts are needed to shift monitoring approaches to better incorporate the gender dimensions of water security, including but not limited to sex-disaggregated data.
- New implementation strategies that demonstrate how gender can be better integrated into water security initiatives at all levels and scales are emerging. Moving from an unaware, harmful, gender-insensitive approach to more transformative practices enables more effective water security outcomes and supports changes toward wider gender equality.

Building on the development of the gendered water security framework and the identification of case studies of gender-sensitive, gender-responsive, and gender-transformative water security practices, this report offers four recommendations to strengthen the integration of water security and gender.

Build internal capacity to address gender equality in water-related international agencies and development partners. This should include roles that are primarily focused on technical aspects of infrastructure design. A critical element to introduce is do no harm strategies, which consider unintentional harm or backlash from any gender-related strategies that may be introduced. Self-assessment processes can be useful tools to reveal staff perspectives and identify gaps.

A woman at a village hand pump to wash her kitchen utensils. A change in mindset and conceptualization of water security to a social one that foregrounds the human dimensions of water security is fundamental to increased consideration of gender equality and inclusion in water security (photo by Amit Verma/ADB).

Address gender equality across institutions responsible for governance and management. These include basins, irrigation, and water and sanitation service delivery across major forums, utilities, government departments, and civil society organizations. Efforts to shift gender parity in these institutions, particularly at senior levels (through measures such as quotas, and recruitment and retention strategies) could influence organizational culture and norms to be more inclusive and would support diversity in leadership roles and decision-making.

Set targets for and monitor progress on gender equality in water security. This will require collaboration across organizations to build on existing efforts to strengthen monitoring of gender equality in Sustainable Development Goal 6 (clean water and sanitation for all) and investment in baseline data collection and analysis. Strengthened monitoring will require new or adapted measurement tools focused on intra-household experiences, meaningful involvement of women, and socially differentiated impacts.

Research and data collection are needed to uncover barriers to women owning land and accessing finance, which may undermine their ability to participate in irrigation, water allocation programs, and integrated water resources management. Similarly, data on the economic and social consequences of women and girls lacking access to improved sanitation and menstrual hygiene facilities underpins increased and targeted investment in these areas of need, which are connected to transforming water-related institutions to being more inclusive.

Promote a shift toward gender-transformative practice in water security initiatives. This involves a focus on participatory and codesign processes with relevant groups, engagement across and between multiple levels (from local to national), and acknowledgment of intersectional opportunities and challenges. It will also be important to improve the knowledge base, collating and sharing transformative best practices, policies, and approaches to promote women's inclusion in water security.

Meaningful involvement of women in water. ADB promotes a shift toward gender-transformative practice in water security initiatives (photo by Amit Verma/ADB).

Glossary

Empowerment is "the processes by which those who have been denied the ability to make choices acquire such an ability" (Kabeer 1999, 437).

Gender equality is understood to be the "equal rights, responsibilities and opportunities of women and men and girls and boys [and individuals of diverse genders]" (Hannan 2001, 1). This definition purposefully adds the inclusion of other genders, recognizing a shift in thinking from binary to diverse genders.

Gender equity is the "fairness of treatment…" [for individuals of all genders] "…according to their respective needs" (UNESCO 2000, 5).

Intersectionality is the acknowledgment that many individuals experience intersectional categories of historical oppression (Crenshaw 1989). As such, intersectionality includes categories like "race, class…sexuality, gender identity, ethnicity, nation, ability, and age" (Collins 2015, 2).

Water includes (i) water supply and sanitation; (ii) water resources; (iii) water for productive uses, including agriculture, industry, and energy; and (iv) water-related hazards, including floods and droughts. Water, sanitation, and hygiene activities can also be identified as water, sanitation, and hygiene (WASH) interventions.

Water security is "the availability of adequate water to ensure safe and affordable water supply, inclusive sanitation for all, improved livelihoods, and healthy ecosystems, with reduced water-related risks toward supporting sustainable and resilient rural–urban economies in the Asia and Pacific region" (ADB 2022, xviii).

A water-secure future. To achieve SDG 6, it is imperative that gender and inclusion are meaningfully incorporated into policies, governance, and management processes (photo by Nozim Kalandarov/ADB).

References

Adelodun, B. et al. 2021. Assessment of Socioeconomic Inequality based on Virus-Contaminated Water Usage in Developing Countries: A Review. *Environmental Research.* 192 (January). DOI: 10.1016/J.ENVRES.2020.110309.

Agol, D. and P. Harvey. 2018. Gender Differences Related to WASH in Schools and Educational Efficiency. *Water Alternatives.* 11 (2). pp. 284–296. https://www.water-alternatives.org/index.php/alldoc/articles/vol11/v11issue2/437-a11-2-4/file.

ASEAN–Australia Smart Cities Trust Fund (AASCTF). 2021. Policy and Practice Recommendations: Towards a Gender Transformative Flood Early Warning System in Baguio City. *AASCTF Policy Brief.* Government of Australia Department of Foreign Affairs and Trade, and Asian Development Bank. https://events.development.asia/system/files/materials/2021/09/202109-aasctf-policy-brief-policy-and-practice-recommendations-gender-transformative-flood-early.pdf.

Asian Development Bank (ADB). 2019. *Strategy 2030 Operational Priority 2: Accelerating Progress in Gender Equality, 2019–2024.* https://www.adb.org/sites/default/files/institutional-document/495956/strategy-2030-op2-gender-equality.pdf.

ADB. 2022. *Strategy 2030 Water Sector Directional Guide—A Water-Secure and Resilient Asia and the Pacific.* https://www.adb.org/sites/default/files/institutional-document/855231/strategy-2030-water-sector-directional-guide.pdf.

Agarwal, B. 2001. Participatory Exclusions, Community Forestry, and Gender: An Analysis for South Asia and a Conceptual Framework. *World Development.* 29 (10). pp. 1623–1648. DOI: 10.1016/S0305-750X(01)00066-3.

Alhassan, S. I., J. K. M. Kuwornu, and Y. B. Osei-Asare. 2019. Gender Dimension of Vulnerability to Climate Change and Variability: Empirical Evidence of Smallholder Farming Households in Ghana. *International Journal of Climate Change Strategies and Management.* 11 (2). pp. 195–214. DOI: 10.1108/IJCCSM-10-2016-0156.

Arana, M. T., A. Quezada, and R. Clements. 2016. How Do Gender Approaches Improve Climate Compatible Development? Lessons from Peru. *Climate & Development Knowledge Network Policy Brief.* May. Cape Town: Climate & Development Knowledge Network. https://cdkn.org/sites/default/files/files/Peru-gender-brief-FINAL.pdf.

Awume, O., R. Patrick, and W. Baijius. 2020. Indigenous Perspectives on Water Security in Saskatchewan, Canada. *Water (Switzerland).* 12 (3). DOI: 10.3390/w12030810.

Baker, T. J. et al. 2015. A Socio-Hydrological Approach for Incorporating Gender into Biophysical Models and Implications for Water Resources Research. *Applied Geography.* 62. pp. 325–338. DOI: 10.1016/j.apgeog.2015.05.008.

Best, J. P. 2019. *(In)visible Women: Representation and Conceptualization of Gender in Water Governance and Management.* Master Thesis. Oregon State University, Corvallis.

Bisung, E. and S. J. Elliott. 2017. Psychosocial Impacts of the Lack of Access to Water and Sanitation in Low- and Middle-Income Countries: A Scoping Review. *Journal of Water and Health*. 15 (1). pp. 17–30. DOI: 10.2166/wh.2016.158.

Brown, S. et al. 2019. *Gender Transformative Early Warning Systems: Experiences from Nepal and Peru*. https://wrd.unwomen.org/sites/default/files/2021-11/GENDER~1_1.PDF.

Bryan, E. and N. Lefore. 2021. Women and Small-Scale Irrigation of Participation and Benefits. *IFPRI Discussion Paper*. No. 02025. DOI: 10.2139/ssrn.3857886.

Bryan, E. and D. Mekonnen. 2023. Does Small-Scale Irrigation Provide a Pathway to Women's Empowerment? Lessons from Northern Ghana. *Journal of Rural Studies*. 97 (January). pp. 474–484. DOI: 10.1016/j.jrurstud.2022.12.035.

Butte, G. et al. 2022. A Framework for Water Security Data Gathering Strategies. *Water (Switzerland)*. 14 (18). DOI: 10.3390/w14182907.

Canipari, R., L. De Santis, and S. Cecconi. 2020. Female Fertility and Environmental Pollution. *International Journal of Environmental Research and Public Health*. 17 (23). 8802. DOI: 10.3390/ijerph17238802.

CARE International. 2022. How Does Irrigation Help Women Be Heard? https://www.care-international.org/stories/how-does-irrigation-help-women-be-heard.

Carrard, N. et al. 2022. The Water, Sanitation and Hygiene Gender Equality Measure (WASH-GEM): Conceptual Foundations and Domains of Change. *Women's Studies International Forum*. 91 (2022). DOI: 10.1016/j.wsif.2022.102563.

Caruso, B. A. et al. 2022. Water, Sanitation, and Women's Empowerment: A Systematic Review and Qualitative Metasynthesis. *PLOS Water*. 1 (6). DOI: 10.1371/journal.pwat.0000026.

Caruso, B.A. et al. 2021. *A Conceptual Framework to Inform National and Global Monitoring of Gender Equality in WASH*. New York: World Health Organization (WHO) and United Nations Children's Fund (UNICEF) Joint Monitoring Programme (JMP) for Water Supply, Sanitation, and Hygiene. https://washdata.org/report/jmp-2021-gender-review-conceptual-framework.

Chua, M. L. et al. 2021. Seasonal and Gender Impacts on Fecal Exposure Trends in an Urban Slum. *Journal of Water and Health*. 19 (6). IWA Publishing. pp. 946–958. DOI: 10.2166/WH.2021.111.

Cleaver, F. 1998. Incentives and Informal Institutions: Gender and the Management of Water. *Agriculture and Human*. 15. pp. 347–360. DOI: https://doi.org/10.1023/A:1007585002325.

Collins, P. H. 2015. Intersectionality's Definitional Dilemmas. *Annual Review of Sociology*. 41. pp. 1–20. DOI: 10.1146/annurev-soc-073014-112142.

Cornwall, A. 2003. Whose Voices? Whose Choices? Reflections on Gender and Participatory Development. *World Development*. 31 (8). pp. 1325–1342. DOI: 10.1016/S0305-750X(03)00086-X.

Crenshaw, Kimberle Williams. 1989. Demarginalizing the Intersection of Race and Sex: A Black Feminist Critique of Antidiscrimination Doctrine, Feminist Theory and Antiracist Politics. *University of Chicago Legal Forum*. (1989). pp. 139–167.

Crow, B. and F. Sultana. 2002. Gender, Class, and Access to Water: Three Cases in a Poor and Crowded Delta. *Society and Natural Resources*. 15 (8). pp. 709–724. DOI: 10.1080/08941920290069308.

Das, P. 2014. Women's Participation in Community-Level Water Governance in Urban India: The Gap Between Motivation and Ability. *World Development.* 64. pp. 206–218. DOI: 10.1016/j.worlddev.2014.05.025.

De Guzman, K. et al. 2023. Drinking Water and the Implications for Gender Equity and Empowerment: A Systematic Review of Qualitative and Quantitative Evidence. *International Journal of Hygiene and Environmental Health.* 247. 114044. DOI: 10.1016/j.ijheh.2022.114044.

De Nys, E. et al. 2013. Empowering Women in Irrigation Management: The Sierra in Peru. *Latin America and Caribbean Region, Environment and Water Resources, Occasional Paper Series.* World Bank. https://documents1.worldbank.org/curated/en/722401468088459634/pdf/768920WP0P14450rrigation0Management.pdf.

Dickin, S. and M. A. Caretta. 2022. Examining Water and Gender Narratives and Realities. *Wiley Interdisciplinary Reviews: Water.* 9 (5). pp. 1–10. DOI: 10.1002/wat2.1602.

Drechsel, P., M. Qadir, and D. Galibourg. 2022. The WHO Guidelines for Safe Wastewater Use in Agriculture: A Review of Implementation Challenges and Possible Solutions in the Global South. *Water 2022.* 14 (6). Multidisciplinary Digital Publishing Institute. p. 864. DOI: 10.3390/W14060864.

Earle, A. and S. Bazilli. 2013. A Gendered Critique of Transboundary Water Management. *Feminist Review.* 103 (1). pp. 99–119. DOI: 10.1057/fr.2012.24.

Elliott, M. et al. 2019. Addressing How Multiple Household Water Sources and Uses Build Water Resilience and Support Sustainable Development. *npj Clean Water.* 2 (1). DOI: 10.1038/s41545-019-0031-4.

Elmendorf, M. and P. Buckles. 1980. *Sociocultural Aspects of Water Supply and Excreta Disposal.* World Bank.

Food and Agriculture Organization of the United Nations (FAO). 2019. Dimitra Clubs in Action. *Dimitra Newsletter: Gender, Rural Women and Development.* No. 30. https://www.fao.org/in-action/kore/publications/publications-details/en/c/1192782/.

FAO. 2021. Water and Gender. https://www.fao.org/land-water/water/watergovernance/water-gender/en/.

Forsey, A. and E. Bessonova. 2020. 5 Ways of Reducing Pollution can Improve Equality for Women. Stockholm Environment Institute. 6 March. https://www.sei.org/featured/5-ways-reducing-pollution-can-improve-equality-for-women/.

Fröhlich, C. et al., eds. 2018. *Water Security Across the Gender Divide.* Springer International Publishing. DOI: 10.1007/978-3-319-64046-4.

Genter, F. et al. 2023. Understanding Household Self-Supply Use and Management Using a Mixed-Methods Approach in Urban Indonesia. *PLOS Water.* 2 (1). e0000070. DOI: 10.1371/journal.pwat.0000070.

Gero, A., T. Chowdhury, and K. Winterford. 2022. *Climate Change Action through Civil Society Programs Partnerships: Research Exploring the Best Practice of Climate Change and Disaster Resilience Integration into Pacific Civil Society Programs.* Prepared by the University of Technology Sydney's Institute for Sustainable Futures (UTS-ISF) for the Australia Pacific Climate Partnership. http://hdl.handle.net/10453/164162.

Girard, A. M. 2014. Stepping into Formal Politics: Women's Engagement in Formal Political Processes in Irrigation in Rural India. *World Development.* 57. pp. 1–18. DOI: 10.1016/j.worlddev.2013.11.010.

Gonzalez, D. et al. 2022. Qualities of Transformative Leaders in WASH: A Study of Gender-Transformative Leadership during the COVID-19 Pandemic. *Frontiers in Water.* 4. DOI: 10.3389/frwa.2022.1050103.

Grant, M. 2017. *Gender Equality and Inclusion in Water Resources Management.* Global Water Partnership Action Piece. Global Water Partnership. http://www.gwp.org/globalassets/global/about-gwp/publications/gender/gender-action-piece.pdf.

Grant, M. and T. Megaw. 2019. *Review of the Implementation of WaterAid's Gender Manual and Facilitated Sessions in Timor-Leste.* UTS-ISF. https://washmatters.wateraid.org/sites/g/files/jkxoof256/files/review-of-the-implementation-of-wateraids-gender-manual-and-facilitated-sessions-in-timor-leste_0.pdf.

Grant, M. et al. 2019. Rural Piped-Water Enterprises in Cambodia: A Pathway to Women's Empowerment? *Water.* 11 (12). pp. 1–18. DOI: 10.3390/w11122541.

Grant, M. et al. 2017. *Female Water Entrepreneurs in Cambodia: Considering Enablers and Barriers to Women's Empowerment.* Enterprise in WASH – Research Report 9. UTS-ISF. https://opus.lib.uts.edu.au/handle/10453/154100.

Grant, M., J. Willetts, and C. Huggett. 2019. *Gender Equality and Goal 6 – The Critical Connection: An Australian Perspective.* Australian Water Partnership. https://waterpartnership.org.au/wp-content/uploads/2019/08/Gender-Equality-and-Goal-6-The-Critical-Connection.pdf.

Grasham, C. F., M. Korzenevica, and K. J. Charles. 2019. On Considering Climate Resilience in Urban Water Security: A Review of the Vulnerability of the Urban Poor in Sub-Saharan Africa. *Wiley Interdisciplinary Reviews: Water.* 6 (3). pp. 1–11. DOI: 10.1002/WAT2.1344.

Halcrow, G. et al. 2010. *Resource Guide: Working Effectively with Women and Men in Water, Sanitation and Hygiene Programs.* International Women's Development Agency and UTS-ISF. http://www.genderinpacificwash.info/system/resources/BAhbBlsHOgZmIjoyMDExLzAxLzI0LzE5LzA0LzQyLzkzMS9XQVNIX2ZsYXNoY2FyZHNfZmluYWw0d2ViLnBkZg/WASH_flashcards_final4web.pdf.

Hannan, C. 2001. Gender Mainstreaming: Strategy for Promoting Gender Equality. https://www.un.org/womenwatch/osagi/pdf/factsheet1.pdf.

Hanrahan, M. and N. Mercer. 2019. Gender and Water Insecurity in a Subarctic Indigenous Community. *Canadian Geographer.* 63 (02). DOI: 10.1111/cag.12508.

Hathi, P. et al. 2021. When Women Eat Last: Discrimination at Home and Women's Mental Health. *PLOS One.* 16 (3 March). pp. 1–22. DOI: 10.1371/journal.pone.0247065.

Howard, E. 2023. Linking Gender, Climate Change and Security in the Pacific Islands Region: A Systematic Review. *Ambio.* 52 (3). pp. 518–533. DOI: 10.1007/s13280-022-01813-0.

iDE. 2020. Design for Developing Countries: Establishing a Methodology for Human-Centered Design in a Challenging Context. https://www.ideglobal.org/story/hcd-toolkit.

Imburgia, L. 2019. Irrigation and Equality: An Integrative Gender-Analytical Approach to Water Governance with Examples from Ethiopia and Argentina. *Water Alternatives.* 12 (3). pp. 571–587. https://www.water-alternatives.org/index.php/alldoc/articles/vol12/v12issue3/543-a12-2-26/file.

Indarti, N. et al. 2019. Women's Involvement in Economic Opportunities in Water, Sanitation and Hygiene (WASH) in Indonesia: Examining Personal Experiences and Potential for Empowerment. *Development Studies Research.* 6 (1). pp. 76–91. DOI: 10.1080/21665095.2019.1604149.

Intergovernmental Panel on Climate Change (IPCC). 2023. *Climate Change 2022 – Impacts, Adaptation and Vulnerability: Working Group II Contribution to the Sixth Assessment Report of the Intergovernmental Panel on Climate Change.* Cambridge University Press. DOI: 10.1017/9781009325844.

International Crops Research Institute for the Semi-Arid Tropics (ICRISAT). 2019. *Gender- and Social-Inclusion Approach in Watershed Project: Insights on Gender Norms and Gender Relations in Parasai-Sindh Watershed, India.* Working Paper. https://oar.icrisat.org/11101/.

International Water Association (IWA). 2022. *IWA Gender Diversity and Water Award.*

IWA. 2014. *An Avoidable Crisis: WASH Human Resource Capacity Gaps in 15 Developing Economies.* https://iwa-network.org/wp-content/uploads/2016/03/1422745887-an-avoidable-crisis-wash-gaps.pdf.

Jagnoor, J. et al. 2019. Exploring the Impact, Response and Preparedness to Water-Related Natural Disasters in the Barisal Division of Bangladesh: A Mixed Methods Study. *BMJ Open.* 9 (4). pp. 1–10. DOI: 10.1136/bmjopen-2018-026459.

Jain, M. et al. 2021. Groundwater Depletion will Reduce Cropping Intensity in India. *Science Advances.* 7 (9). pp. 1–10. DOI: 10.1126/sciadv.abd2849.

Jaren, L. S., R. S. Leya, and M. S. Mondal. 2022. Investigation of Gender-Differentiated Impacts of Water Poverty on Different Livelihood Groups in Peri-Urban Areas around Dhaka, Bangladesh. *Water.* 14 (7). DOI: 10.3390/w14071167.

Jepson, W. E., and E. Vandewalle. 2016. Household Water Insecurity in the Global North: A Study of Rural and Periurban Settlements on the Texas–Mexico Border. *Professional Geographer.* 68 (1). pp. 66–81. DOI: 10.1080/00330124.2015.1028324.

Jepson, W. E. et al. 2017a. Advancing Human Capabilities for Water Security: A Relational Approach. *Water Security.* 1. pp. 46–52. DOI: 10.1016/j.wasec.2017.07.001.

Jepson, W. E. et al. 2017b. Progress in Household Water Insecurity Metrics: A Cross-Disciplinary Approach. *Wiley Interdisciplinary Reviews: Water.* 4 (3). pp. 1–21. DOI: 10.1002/WAT2.1214.

JMP. 2023. *Progress on Household Drinking Water, Sanitation and Hygiene 2000–2022: Special Focus on Gender.* UNICEF and WHO. https://washdata.org/reports/jmp-2023-wash-households.

JMP. 2021. *Progress on Household Drinking Water, Sanitation and Hygiene 2000–2022: Five Years Into the SDGs.* UNICEF and WHO. https://washdata.org/reports/jmp-2021-wash-households.

JMP. 2019. *Progress on Household Drinking Water, Sanitation and Hygiene 2000–2017: Special Focus on Inequalities.* UNICEF and WHO. https://washdata.org/reports/progress-household-drinking-water-sanitation-and-hygiene-2000-2017-special-focus.

Kabeer, N. 1999. Resources, Agency, Achievements: Reflections on the Measurement of Women's Empowerment. *Development and Change.* 30 (3). pp. 435–464. DOI: 10.1111/1467-7660.00125.

Kabeer, N. 2018. Randomized Control Trials and Qualitative Evaluations of a Multifaceted Program for Women in Extreme Poverty: Empirical Findings and Methodological Reflections. *Journal of Human Development and Capabilities.* 20 (2). pp. 197–217. DOI: 10.1080/19452829.2018.1536696.

Kayser, G. L. et al. 2019. Water, Sanitation and Hygiene: Measuring Gender Equality and Empowerment. *Bulletin of the World Health Organization.* 97 (6). pp. 438–440. DOI: 10.2471/BLT.18.223305.

Kelm, U. et al. 2017. *Is Wastewater a She? Linking SDG 6.3 (wastewater) and SDG 5 (gender): Conclusions and Recommendations.* Presentation prepared for World Water Week 2017. Stockholm. 27 August–1 September. https://programme.worldwaterweek.org/Content/ProposalResources/PDF/2017/pdf-2017-6782-1-Is%20wastewater%20a%20She_final%20recommendations.pptx.pdf.

Kumar, A., M. Grant, and J. Willetts. 2021. *Inclusive Water, Sanitation and Hygiene (WASH) Workplaces: Guidance for the WASH Sector.* UTS-ISF. https://multisitestaticcontent.uts.edu.au/wp-content/uploads/sites/57/2021/11/04190001/ISF-UTS_2021_Guidance-on-Inclusive-WASH-workplaces.pdf.

Leahy, C. et al. 2017. Transforming Gender Relations through Water, Sanitation, and Hygiene Programming and Monitoring in Vietnam. *Gender and Development.* 25 (2). pp. 283–301. DOI: 10.1080/13552074.2017.1331530.

Leder, K. et al. 2021. Study Design, Rationale and Methods of the Revitalising Informal Settlements and their Environments (RISE) Study: A Cluster Randomised Controlled Trial to Evaluate Environmental and Human Health Impacts of a Water-Sensitive Intervention in Informal Settlements in Indonesia and Fiji. *BMJ Open.* 11 (1). DOI: 10.1136/bmjopen-2020-042850.

Lee, L. et al. 2018. *Gender and WASH Monitoring Tool.* Plan International Australia. https://www.plan.org.au/wp-content/uploads/2020/08/gender-and-wash-monitoring-tool-2018.pdf.

Leya, R. S. 2019. *Gender Inclusive Water Security Framework for Peri-Urban Area in the Context of Climate Change.* Bangladesh University of Engineering and Technology. http://lib.buet.ac.bd:8080/xmlui/handle/123456789/5791.

MacArthur, J. and S. Koh. 2022. *Exploring Gendered Experiences within iDE Cambodia's SMSU3 WASH Program: Photo-Stories.* UTS-ISF. https://opus.lib.uts.edu.au/handle/10453/159765.

MacArthur, J. et al. 2023. Gender Equality Approaches in Water, Sanitation, and Hygiene Programs: Towards Gender-Transformative Practice. *Frontiers in Water.* 5. DOI: 10.3389/frwa.2023.1090002.

MacArthur, J. et al. 2022a. *Climate Impacts on Gender Equality in WASH: A Photovoice Assessment.* UTS-ISF Learning Brief. UTS-ISF. https://multisitestaticcontent.uts.edu.au/wp-content/uploads/sites/57/2023/01/12032538/Climate-impacts-on-gender-equality-in-WASH_-A-photovoice-assessment.pdf.

MacArthur, J. et al. 2022b. Gender-Transformative Approaches in International Development: A Brief History and Five Uniting Principles. *Women's Studies International Forum.* 95 (2022). DOI: 10.1016/j.wsif.2022.102635.

Marks, S. J. et al. 2016. An Impact Evaluation of Multiple-Use Water Services in the Morogoro Region of Tanzania. Duebendorf, Switzerland and Miami: Sandec Department at Eawag and the Global Water for Sustainability Program at Florida International University. https://www.eawag.ch/fileadmin/Domain1/Abteilungen/sandec/schwerpunkte/WST/impact_evaluation_water_services_morogoro.pdf.

McOmber, C., C. Audia, and F. Crowley. 2019. Building Resilience by Challenging Social Norms: Integrating a Transformative Approach within the BRACED Consortia. *Disasters.* 43 (S3). pp. S271–S294. DOI: 10.1111/disa.12341.

Megaw, T. and K. Winterford. 2020. A Gender-Transformative Social Accountability Model. *Gender Transformative Social Accountability*. Working Paper 2. UTS-ISF. https://multisite staticcontent.uts.edu.au/wp-content/uploads/sites/57/2020/11/30110628/GTSA-WP-No.2-V5-web.pdf.

Meierhofer, R. et al. 2022. Water Carrying in Hills of Nepal—Associations with Women's Musculoskeletal Disorders, Uterine Prolapse, and Spontaneous Abortions. *PLOS One.* 17 (6 June). pp. 1–23. DOI: 10.1371/journal.pone.0269926.

Melbourne Water. 2022. *Gender Equality Action Plan 2022–2025.* https://www.melbournewater.com.au/media/20296/download.

Merkle, O., U. Allakulov, and D. Gonzalez. 2022. Sextortion in Access to WASH Services in Selected Regions of Bangladesh. *United Nations University, Maastricht Economic and Social Research Institute on Innovation and Technology Working Paper Series*. No. 2022-022. https://cris.maastrichtuniversity.nl/en/publications/sextortion-in-access-to-wash-services-in-selected-regions-of-bang.

Metropolitan Waterworks and Sewerage System (MWSS). 2023. Gender and Development Database. https://mwss.gov.ph/human-resources-matters/gender-and-development/

Moayed, K. et al. 2020. Addressing Gender-Inclusive Public Health in Climate Change Response: Women in Water-Related Disaster. *Journal of Climate Change Research.* 11 (5–1). pp. 427–436. DOI: 10.15531/ksccr.2020.11.5.427.

Mommen, B., K. Humphries-Waa, and S. Gwavuya. 2017. Does Women's Participation in Water Committees Affect Management and Water System Performance in Rural Vanuatu? *Waterlines.* 36 (3). pp. 216–232. DOI: 10.3362/1756-3488.16-00026.

Myrttinen, H. 2017. The Complex Ties that Bind: Gendered Agency and Expectations in Conflict and Climate Change-Related Migration. *Global Policy.* 8 (February). pp. 48–54. DOI: 10.1111/1758-5899.12402.

Narain, V. and D. Roth, eds. 2002. *Water Security, Conflict and Cooperation in Asia Peri-Urban South: Flows across Boundaries*. DOI: 10.1007/978-3-030-79035-6.

Nunbogu, A. M. and S. J. Elliott. 2022. Characterizing Gender-Based Violence in the Context of Water, Sanitation, and Hygiene: A Scoping Review of Evidence in Low- and Middle-Income Countries. *Water Security.* 15 (February). DOI: 10.1016/j.wasec.2022.100113.

Nussbaum, M. C. 2000. *Women and Human Development: The Capabilities Approach.* Cambridge University Press. DOI: 10.1017/CBO9780511841286.

Nyukuri, E., L. O. Naess, and L. Schipper. 2016. How Do Gender Approaches Improve Climate Compatible Development? Lessons from Kenya. *Climate & Development Knowledge Network Policy Brief.* May. Climate & Development Knowledge Network. https://africaportal.org/wp-content/uploads/2023/05/Kenya-gender-brief-FINAL.pdf.

Octavianti, T. and C. Staddon. 2021. A Review of 80 Assessment Tools Measuring Water Security. *Wiley Interdisciplinary Reviews: Water.* 8 (3). pp. 1–24. DOI: 10.1002/wat2.1516.

O'Reilly, K. 2004. Developing Contradictions: Women's Participation as a Site of Struggle within an Indian NGO. *Professional Geographer.* 56 (2). pp. 174–184. DOI: 10.1111/j.0033-0124.2004.05602003.x.

Padmaja, R., L. B. Stephanie, and G. Kaushal. 2018. Gender- and Social- Inclusion Approach in Watershed Projects: Insights on Gender Norms and Gender Relations in Parasai-Sindh Watershed, India. DOI: 10.13140/RG.2.2.30699.28964.

Pearl-Martinez, R. 2017. Financing Women Farmers. *OXFAM Briefing Paper*. https://oxfamilibrary.openrepository.com/bitstream/handle/10546/620352/bp-financing-women-farmers-131017-en.pdf;jsessionid=8587FE096E66546C91751B62CC59D393?sequence=10.

Pouramin, P., N. Nagabhatla, and M. Miletto. 2020. A Systematic Review of Water and Gender Interlinkages: Assessing the Intersection with Health. *Frontiers in Water.* 2 (April). DOI: 10.3389/frwa.2020.00006.

Poutiainen, P. and A. Mills. 2014. Mainstreaming Gender in the Irrigation Development Support Programme—Case Study, Zambia. *Agriculture and Environmental Services Department Notes.* World Bank. https://documents1.worldbank.org/curated/en/474461468350180137/pdf/849500BRI0AES000Box382145B00PUBLIC0.pdf.

Rautanen, S. L. and U. Baaniya. 2008. Technical Work of Women in Nepal's Rural Water Supply and Sanitation. *Water International.* 33 (2). pp. 202–213. DOI: 10.1080/02508060802027687.

Robeyns, I. 2005. The Capability Approach: A Theoretical Survey. *Journal of Human Development.* 6 (1). pp. 93–117. DOI: 10.1080/146498805200034266.

Sehring J., R. ter Horst, and M. Zwarteveen, eds. 2022 *Gender Dynamics in Transboundary Water Governance: Feminist Perspectives on Water Conflict and Cooperation*. Routledge. DOI: 10.4324/9781003198918-4.

Sinharoy, S. et al. 2023. The Agency, Resources, and Institutional Structures for Sanitation-Related Empowerment (ARISE) Scales: Development and Validation of Measures of Women's Empowerment in Urban Sanitation for Low- and Middle-Income Countries. *World Development.* 164. DOI: 10.2139/ssrn.4149004.

Soeters, S. et al. 2019. Intersectionality: Ask the Other Question. *Water for Women: Gender in WASH.* Conversational Article 2. UTS-ISF. https://www.waterforwomenfund.org/en/resources/en/resourcesGeneral/news/1903ISF-UTS_2019_IntersectionalityArticle2_GenderinWASH_WaterforWomen.pdf.

Soeters, S. et al. 2020. *Gender Equality and Women in Water, Sanitation and Hygiene (WASH) Enterprises in Cambodia: A Synthesis of Recent Studies.* UTS-ISF. https://www.waterforwomenfund.org/en/learning-and-resources/resources/KL/2007-Cambodia-Enterprise-Synthesis.pdf.

Sogani, R. 2016. How Do Gender Approaches Improve Climate Compatible Development? Lessons from India. *Climate & Development Knowledge Network Policy Brief.* May. Climate & Development Knowledge Network. https://wrd.unwomen.org/practice/resources/how-do-gender-approaches-improve-climate-compatible-development-lessons-india.

Stevenson, E. G. J. et al. 2016. Community Water Improvement, Household Water Insecurity, and Women's Psychological Distress: An Intervention and Control Study in Ethiopia. *PLOS One.* 11 (4). DOI: 10.1371/journal.pone.0153432.

Swaraj, A. and B. Maheshwari. 2022. Understanding the Role of Groundwater in the Lives of Rural Women in India. *World Water Policy.* 8 (2). pp. 116–131. DOI: 10.1002/wwp2.12085.

Sultana, F. 2009. Fluid Lives: Subjectivities, Gender and Water in Rural Bangladesh. *Gender, Place and Culture.* 16 (4). pp. 427–444. DOI: 10.1080/09663690903003942.

Tallman, P. S. et al. 2022. Water Insecurity and Gender-Based Violence: A Global Review of the Evidence. *Wiley Interdisciplinary Reviews: Water.* 10 (1). pp. 1–19. DOI: 10.1002/wat2.1619.

Tandon, I. et al. 2022. Urban Water Insecurity and its Gendered Impacts: On the Gaps in Climate Change Adaptation and Sustainable Development Goals. *Climate and Development.* DOI: 10.1080/17565529.2022.2051418.

Tantoh, H. B. et al. 2021. Gender Roles, Implications for Water, Land, and Food Security in a Changing Climate: A Systematic Review. *Frontiers in Sustainable Food Systems.* 5 (July). DOI: 10.3389/fsufs.2021.707835.

Taron, A., P. Drechsel, and S. Gebrezgabher. 2021. *Gender Dimensions of Solid and Liquid Waste Management for Reuse in Agriculture in Asia and Africa.* Resource Recovery & Refuse Series 21. International Water Management Institute and CGIAR Research Program on Water, Land and Ecosystems. DOI: 10.5337/2021.223. https://www.iwmi.cgiar.org/Publications/wle/rrr/resource_recovery_and_reuse-series_21.pdf.

Tesfaye, Y. et al. 2020. How Do Rural Ethiopians Rate the Severity of Water Insecurity Scale Items? Implications for Water Insecurity Measurement and Interventions. *Human Organization.* 79 (2). pp. 95–106. DOI: 10.17730/1938-3525.79.2.95.

Theis, S. et al. 2016. *Integrating Gender into Small-Scale Irrigation.* Feed the Future Innovation Lab for Small Scale Irrigation Project Notes 2. International Food Policy Research Institute and International Water Management Institute. http://ebrary.ifpri.org/cdm/ref/collection/p15738coll2/id/131549.

Thompson, J. et al. 2001. Drawers of Water II: Assessing Change in Domestic Water Use in East Africa. *Waterlines.* 22 (1). DOI: 10.3362/0262-8104.2003.039.

Thompson, J. A. 2016. Intersectionality and Water: How Social Relations Intersect with Ecological Difference. *Gender, Place & Culture* 23 (9). pp. 1286–1301. DOI: 10.1080/0966369X.2016.1160038.

Truelove, Y. 2019. Rethinking Water Insecurity, Inequality and Infrastructure through an Embodied Urban Political Ecology. *Wiley Interdisciplinary Reviews: Water.* 6 (3). pp. 1–7. DOI: 10.1002/WAT2.1342.

UN Women. 2018. *Gender Equality in the 2030 Agenda: Gender-Responsive Water and Sanitation Systems.* Issue Brief. https://www.unwomen.org/sites/default/files/Headquarters/Attachments/Sections/Library/Publications/2018/Issue-brief-Gender-responsive-water-and-sanitation-systems-en.pdf.

United Nations Development Programme (UNDP) – Stockholm International Water Institute Water Governance Facility. 2017. Women and Corruption in the Water Sector: Theories and Experiences from Johannesburg and Bogotá. *Water Governance Facility Report.* No. 8. https://siwi.org/publications/wgf-report-8-women-corruption-water-sector-theories-experiences-johannesburg-bogota/.

United Nations Educational, Scientific and Cultural Organization (UNESCO). 2000. Gender Equality and Equity: A Summary Review of UNESCO's Accomplishments since the Fourth World Conference on Women (Beijing 1995). https://unesdoc.unesco.org/ark:/48223/pf0000121145.

Ungureanu, N., V. Vlăduț, and G. Voicu. 2020. Water Scarcity and Wastewater Reuse in Crop Irrigation. *Sustainability.* 12 (21). DOI: 10.3390/SU12219055.

United Nations Children's Fund (UNICEF). 2016. Collecting Water is often a Colossal Waste of Time for Women and Girls. *Press Release.* 29 August. https://www.unicef.org/press-releases/unicef-collecting-water-often-colossal-waste-time-women-and-girls.

UNICEF, WaterAid, and Water and Sanitation for the Urban Poor. 2018. *Female-Friendly Public and Community Toilets: A Guide for Planners and Decision Makers.* WaterAid. https://washmatters.wateraid.org/sites/g/files/jkxoof256/files/female-friendly-public-and-community-toilets-a-guide.pdf.

van Hoeve, E. and B. van Koppen. 2005. Beyond Fetching Water for Livestock: A Gendered Sustainable Livelihood Framework to Assess Livestock-Water Productivity. In D. Peden, ed. 2005. *Nile Basin Water Productivity: Developing a Shared Vision for Livestock Production.* Proceedings prepared for a workshop of the CGIAR Challenge Program on Water & Food. Kampala, Uganda, 5–8 September.

van Houweling, E. et al. 2012. The Role of Productive Water Use in Women's Livelihoods: Evidence from Rural Senegal. *Water Alternatives.* 5 (3): pp. 658–677. https://www.water-alternatives.org/index.php/volume5/v5issue3/191-a5-3-7/file.

van Koppen, B. et al. 2021. Gender Equality and Social Inclusion in Community-led Multiple Use Water Services in Nepal. *International Water Management Institute Working Papers.* No. 203. International Water Management Institute. DOI: 10.5337/2022.200. http://www.iwmi.cgiar.org/Publications/Working_Papers/working/wor203.pdf.

Varua, M. E. et al. 2018. Groundwater Management and Gender Inequalities: The Case of Two Watersheds in Rural India. *Groundwater for Sustainable Development.* 6 (March). pp. 93–100. DOI: 10.1016/j.gsd.2017.11.007.

Water for Women. 2022a. *Civil Society Engagement with the Private Sector for Inclusive WASH: Insights from Water for Women.* https://www.waterforwomenfund.org/en/learning-and-resources/resources/KL/Publications/ETPSII-WASH-Learning-Brief-FINAL.pdf.

Water for Women. 2022b. *Partnerships for Transformation: Guidance for WASH and Rights Holder Organisations.* https://www.waterforwomenfund.org/en/learning-and-resources/resources/KL/Publications/Water-for-Women-TT-Partnerships-for-Transformation-Guidance-for-WASH-and-RHOs-web.pdf.

van Wijk-Sijbesma, C. 1997. *Gender in Water Resources Management, Water Supply and Sanitation: Roles and Realities Revisited.* IRC International Water and Sanitation Centre. https://www.ircwash.org/sites/default/files/Wijk-1998-GenderTP33-text.pdf.

Winter, J. C., G. L. Darmstadt, and J. Davis. 2021. The Role of Piped Water Supplies in Advancing Health, Economic Development, and Gender Equality in Rural Communities. *Social Science and Medicine.* 270. DOI: 10.1016/j.socscimed.2020.113599.

World Bank. 2019. *Women in Water Utilities: Breaking Barriers*.

World Economic Forum. 2022. Global Gender Gap Report 2022. *Insight Report*. https://www3.weforum.org/docs/WEF_GGGR_2022.pdf.

Worsham, K. 2021. Leadership for SDG 6.2: Is Diversity Missing? *Environmental Health Insights*. 15. DOI: 10.1177/11786302211031846.

Young, S. L. et al. 2019. The Household Water InSecurity Experiences (HWISE) Scale: Development and Validation of a Household Water Insecurity Measure for Low-Income and Middle-Income Countries. *BMJ Global Health*. 4 (5). DOI: 10.1136/bmjgh-2019-001750.

Young, S. L. et al. 2021a. Perspective: The Importance of Water Security for Ensuring Food Security, Good Nutrition, and Well-Being. *Advances in Nutrition*. 12 (4). pp. 1058–1073. DOI: 10.1093/advances/nmab003.

Young, S. L. et al. 2021b. The Individual Water Insecurity Experiences (IWISE) Scale: Reliability, Equivalence and Validity of an Individual-Level Measure of Water Security. *BMJ Global Health*. 6 (10). pp. 1–14. DOI: 10.1136/bmjgh-2021-006460.

www.ingramcontent.com/pod-product-compliance
Lightning Source LLC
LaVergne TN
LVHW070943160826
845679LV00022B/1902

* 9 7 8 9 2 9 2 7 0 9 2 6 6 *